TO THE FRONT

Grandfathers' Stories in the Cause of Freedom

Also by Michael M. Van Ness

General in Command: The Life of Major General John B. Anderson from Iowa Farm to Command of the Largest Combat Corps in World War Two

To the Front

Grandfathers' Stories in the
Cause of Freedom

BY MICHAEL M. VAN NESS, M.D.

MODERN MEMOIRS, INC.
Amherst, Massachusetts

TO THE FRONT: GRANDFATHERS' STORIES IN THE CAUSE OF FREEDOM
© 2022 Michael M. Van Ness, M.D.

The thoughts, reflections, and opinions expressed in this book are those of the author, based upon his personal recollections and research. In some instances, invented dialog has been inserted as a literary device, drawing on the author's recollections of his lived experiences and his good-faith impressions of people mentioned and described herein. The author takes full and sole responsibility for all of the contents, including text and images, and regrets any aspect of the content that might be construed as injurious to a party mentioned, implied, or referred to. Genealogy chart is based on professional research conducted by Liz Sonnenberg; the discovery of additional sources or interpretations may affect the chart.

Cover design by Nicole Miller
Front cover: Main image from cover of a 1943 edition of *102d Infantry Division Magazine,* courtesy of Albert Love Enterprises, Atlanta, Georgia
Front cover: Ozarks 102nd Infantry Division patch created in 1942
Back cover: Photo courtesy of U.S. Army Signal Corps
Printed and bound in the U.S.A.
Softcover edition ISBN: 978-0-9997705-4-2

Modern Memoirs, Inc.
495 West Street, Suite 1C
Amherst, Massachusetts 01002
413-253-2353
www.modernmemoirs.com

This book is dedicated to my father,
Captain Harper Elliott Van Ness, Jr.,
U.S. Naval Academy, Class of 1943,
destroyer officer, naval aviator,
"steely-eyed" missile man.

DECK LOG ENTRY USS TWIGGS (DD591)
SUNDAY, 1 JAN. 1945

In Kossol Roads we start this book,
With seventy five fathoms to the starboard hook.
Eleven fathoms of water the hand lead reads.
With number one boiler supplying our needs.
Babelthuap of the Palaus looms in sight.
The ship is darker than the very night.
Readiness condition four is set throughout.
The watch is alert, no one else is about.
Various units of the allied fleet,
Lie on all sides where the eye will meet.
Of battleships, cruisers, and carriers untold,
This ship is but one of a hundredfold.

Robert G. Wiltgen, Lt.(jg) USNR

(Bob was killed in action, Okinawa, June 16, 1945)

From the USS Twiggs *Archives, New Year's Day 1945. It is customary for the first deck log entry of the new year to be written in verse, composed by the OD of the mid watch (0000-0400). When Lieutenant Robert "Bob" Wiltgen wrote the following,* Twiggs *was anchored in Ulithi Atoll as part of the Task Force assembling to invade the Philippines at Lingayen Gulf, 10 days later. Just six months later, Lt. Wiltgen was killed in action.*

CONTENTS

LIST OF MAPS

NOTABLE ALLIED LEADERS AND
OFFICER WARTIME RANKS

United States of America

President
  Franklin D. Roosevelt
  Harry S. Truman

United Kingdom

Prime Minister
  Winston S. Churchill

American Army

General of the Army (5-Star)
  George C. Marshall
  Dwight D. Eisenhower

General (4-Star)
  Omar Bradley
  Malin Craig, Sr.

Lieutenant General (3-Star)
  Courtney Hodges
  George Patton
  William H. Simpson
  Carl "Tooey" Spaatz

Major General (2-Star)
  John B. Anderson
  "Lightning Joe" Collins
  Norman Cota
  James Gavin
  Troy Middleton
  Matthew Ridgway

Brigadier General (1-Star)
  Antony McAuliffe

Colonel (O-6)
  Armand Hopkins

Lieutenant Colonel (O-5)
  Malin Craig, Jr.

Major (O-4)
  Aaron Cohn

Captain (O-3)
  John Starkey (WWI)

First Lieutenant (O-2)
  Joe Swing (WWI)

Second Lieutenant (O-1)
  Bill Houghton

American Navy

Fleet Admiral (5-Star)
  Ernest King

Admiral (4-Star)
  Chester W. Nimitz

Vice Admiral (3-Star)
  William Halsey, Jr.

Rear Admiral (2-Star)
  Marc A. Mitscher
  Raymond A. Spruance

Commodore

Captain (O-6)
  Daniel Gallery

Commander (O-5)

Lieutenant Commander (O-4)
  Maxwell Leslie

Lieutenant (O-3)
  Harper "Smiling Jack" Van Ness

Lieutenant (jg)
  Paul Holmberg
  Robert Wiltgen

Ensign
  Humphrey H. Cordes

British Army

Field Marshall (5-Star)
  Bernard L. Montgomery

General (4-Star)
  Miles Dempsey

MAJOR GENERAL JOHN B. ANDERSON'S ROLES WITHIN THE COMMAND STRUCTURE OF THE U.S. ARMY IN WORLD WAR TWO

President of the United States Franklin D. Roosevelt

Secretary of War Henry L. Stimson

Army Chief of Staff General George C. Marshall

Supreme Allied Commander General Dwight D. Eisenhower
Commanded 6th (Devers) and 12th (Bradley) Army groups

12th Army Group Commander General Omar N. Bradley
Commanded First, Third, and Ninth armies

Ninth Army Commander Lieutenant General William H. Simpson
Commanded XII, XVI, and XIX Corps
Army contained 100,000–150,000 soldiers

XVI Corps Commander Major General John B. Anderson
Commanded 8th Armored and 30th and 79th Infantry divisions for
    Rhine Crossing in 1945
Corps contained 20,000–45,000 soldiers

102nd Infantry Division Major General John B. Anderson
Commanded 102nd Infantry division at its inception in 1942
Consisted of the 405th, 406th, and 407th Infantry regiments
Division contained 10,000–15,000 soldiers

* * *

Infantry Regiments
Commanded by a colonel
Consisted of three or more battalions
Contained 2,000–5,000 soldiers

Battalions
Commanded by a lieutenant colonel or major
Consisted of three to five infantry companies
Contained 100–1,000 soldiers

Companies or Batteries
Commanded by a captain or first lieutenant
Contained 100–200 soldiers

Platoons
Commanded by a second lieutenant
Contained 18–50 soldiers

ANCESTOR CHART OF MICHAEL M. VAN NESS

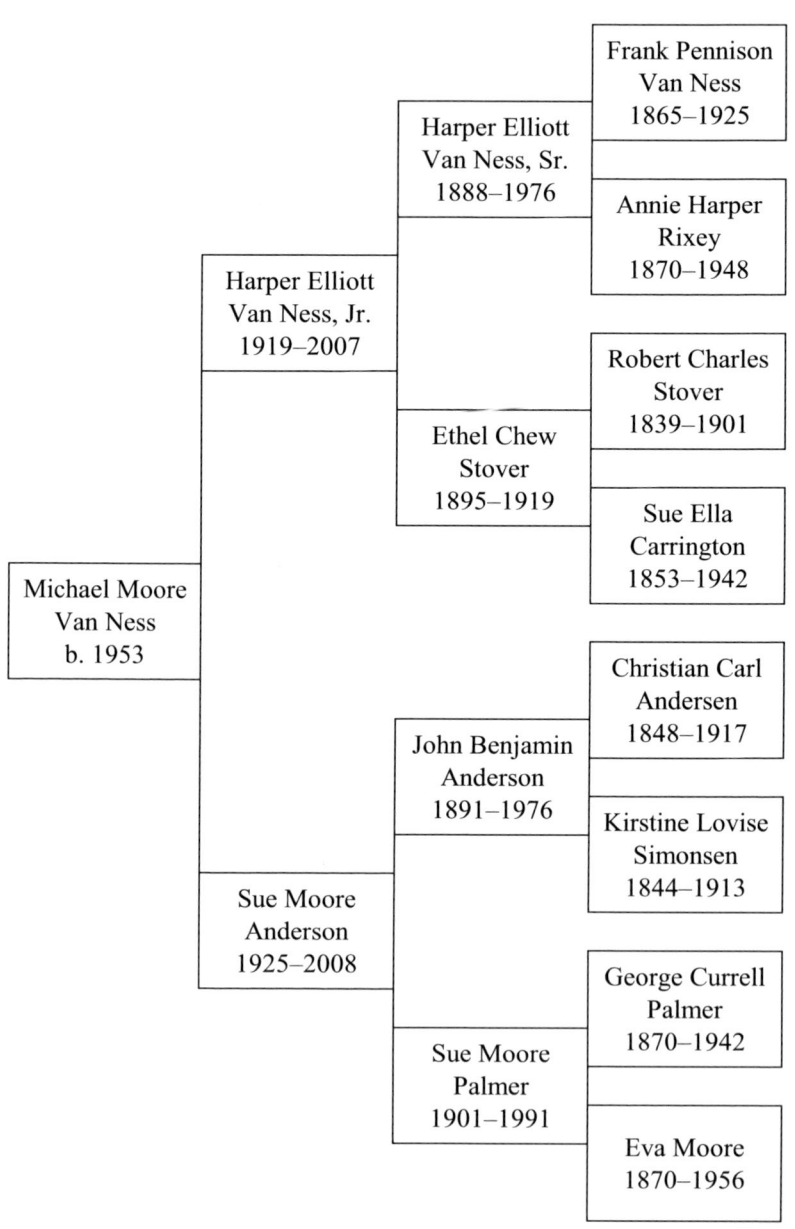

FOREWORD BY THOMAS J. HAMILTON

The cost of freedom is always high, but Americans have always paid it. And one path we shall never choose, and that is the path of surrender, or submission.

—John F. Kennedy

With these brave words, President John F. Kennedy addressed America during the Cuban Missile Crisis of October 1962 as the world stood for the first time on the brink of a possible global thermonuclear war. President Kennedy defined America as a proud people willing to pay whatever price necessary to preserve our freedom. Surrender and submission were not options. Those words proved true in 1962, and let's hope that they still prove true today.

In 1962 Kennedy addressed a nation of mature adults—acquainted with the crippling human cost of defending freedom. Our "greatest generation"—by then approaching middle age—was all too familiar with the extortionate price that freedom so often demands. Their war, World War Two, was a conventional, kinetic war inflicting more than 50 million casualties worldwide before it ended so emphatically with the surrender of Japan. America alone lost some 400,000 courageous souls in this conflict.

The war that threatened America in 1962 was a nuclear war, with far more potential still for devastating consequences. But Kennedy—and the America of 1962—were up to the challenge. Surrender, capitulation, or submission simply were not on the table.

Today America faces a third type of war which is more insidious than either conventional or nuclear war because it is covert: a war of subversion. The ancient Chinese military philosopher Sun Tzu touted the advantages of subversion—conquering your enemy without suffering battlefield casualties.

And international communism—beginning with the Soviet Union's propaganda campaign in the 1920s—long has waged a war of subversion against America. And that war now has reached critical mass. Average Americans are beginning to appreciate that something is *terribly wrong* with our government at every level, its policies, and most importantly its attitude to *We, the People* to whom—at least in theory—government in America remains perpetually accountable.

As President Kennedy fearlessly intoned, patriotic America always pays the price of freedom, regardless of the challenge, regardless of the war. *That is who we are.* This book authentically captures a slice of America and American culture from an era in which patriotism and sacrifice were accepted as essential to our inherent and God-given identity as a free people with a special destiny, with surrender and submission never being options. It relates the tremendous (and ultimately unknowable) sacrifices that our greatest generation made during World War Two to protect our freedom from Imperial Japan and Nazi Germany, unquestionably among the most evil regimes ever to threaten world order.

I have known the author since 1964 when we were schoolboys together at the Landon School for Boys in Bethesda, Maryland. His grandfather, John B. Anderson, was a graduate of West Point, an Army general, and a veteran of both the First and Second World Wars. I also knew his father, Harper E. Van Ness, Jr., a 1943 graduate of the U.S. Naval Academy who served aboard a destroyer, USS *Lansdale*, on convoy duty for two years in the icy North Atlantic before completing flight training for duty flying Hellcats against Japan. My late father, Colonel Lloyd W. Hamilton, was a B-24 Liberator pilot during World War Two who flew 50 missions against Nazi Germany and occupied territories, receiving the Distinguished Flying Cross and the Air Medal for his service.

When I asked him once what it was like to fly combat missions during World War Two, he replied tersely that it was worse than anything that I could possibly imagine. End of conversation. Millions of American families share a similar lineage.

That our past generations sacrificed more than we can ever know to win and preserve our freedom should surprise no one. George Washington said as much about his Continental Army that won our independence from Great Britain: posterity would never appreciate what they had endured. But by reaching back to capture the ethos and culture of traditional, patriotic America that this book recounts so vividly, readers will be better prepared to confront a greater existential challenge to our liberty and freedom than either enemy could pose during World War Two.

—*Mr. Thomas J. Hamilton*

ACKNOWLEDGEMENTS

I would be remiss if I did not acknowledge Megan St. Marie and her colleagues Ali de Groot, Nicole Miller, and Liz Sonnenberg at Modern Memoirs, Inc. Their expertise, attention to detail, and encouragement guided the evolution of the book from a random collection of stories into a coherent work.

My thanks to my brother Elliott Scott Van Ness for his critical reading of the text.

I am grateful to my friends at the Landon School for Boys for photographs and details of their fathers' careers.

I am also grateful to the Cordes family, Ms. Gail Cohn, Dr. Sarkis Chobanian, and Dr. Walter Soduk for allowing me to share their family stories with you.

My wife Sandra Soni Van Ness endured hours of "writer's wife widowhood" as I focused on the manuscript. I thank her for her patience and never-ending encouragement.

While I thank everyone mentioned above for their help, I affirm that all errors, omissions, and oversights are mine alone.

INTRODUCTION

I grew up in a military family. My maternal grandfather, Major General John B. Anderson (née Andersen, changed upon arrival at U.S. Military Academy to "Anderson"), graduated from West Point in 1914. My father, Captain Harper E. Van Ness, Jr., was a U.S. Naval Academy graduate, Class of 1943. My father's father, Harper Elliott Van Ness, was an Army sergeant in World War One. Farther back, my maternal grandfather's brother Nels Andersen fought in the Spanish–American War. My maternal grandmother's grandfather John Palmer joined the Confederate Army in 1862. John Palmer's uncle Samuel Bell was an officer in the Tennessee Volunteer Cavalry and fought in the Mexican–American War, 1846–48.

Growing up, I was surrounded by the stories of these men. Artifacts of their service lay about in the living room, while others were hidden in drawers. In closets hung uniforms stored with dress swords and field boots. Displayed were medals and photographs of people and places from around the world. Some of my first memories are the roar of jet planes, the chatter of a .50 caliber machine gun, and the shrill sound of a bosun's whistle.

It was natural for me to think I would follow in their footsteps. I was a good student and athlete. When my father was stationed at the Pentagon, he would often take me and my two older brothers to events at the nearby Naval Academy—pistol matches, wrestling matches, and football games. I always enjoyed our outings and could see myself wearing Navy blue and gold.

But there was a problem, I was near-sighted. In those days, the need for glasses disqualified one from an Academy appointment. There was another problem. My father was fearful of financial insecurity. In fact, he was obsessed by the fact that he was raised poor in Mexico, Missouri. Although financially secure now as a Naval officer, he was never going to make the kind of money folks in Washington, D.C. did. From an early age I, too, was very conscious that our family circumstances resembled those described by David McCullough in his book *The Pioneers: The Heroic Story of the Settlers Who Brought the American Ideal West.* We were poorer than our neighbors who had not been in the military, but I never felt inferior to them. Naturally, I felt deep pride in my family's military service and tried to carry myself with the dignity befitting a military brat who one day might be an officer and a gentleman.

During my early years, frequent trips to Bethesda Naval Hospital were common, usually for treatment of minor athletic injuries. Those visits got me thinking, "If I can't go to the Naval Academy, maybe I'll be a Navy doctor." My father warmed to the idea; nothing wrong having his son as a Navy doctor. And near-sightedness was not an obstacle to a career in Navy medicine.

Childhood dreams of a Navy medical career were reinforced by a week-long hospitalization at Bethesda for knee surgery in the autumn of 1966. I was thirteen, placed in an open ward with wounded enlisted men from the Vietnam War. They doted on me, the son of a senior officer and his comely wife. I relished the attention and admired the men and their stories of combat and service life. I could see myself a part of this world, a world of courage, adventure, and sacrifice. I liked the sights and smells of clean sheets, antiseptic, and starched uniforms.

My childhood dreams evaporated in the heat of the anti-war movement of the 1960s. The threat of the draft changed the attitude of many of my peers, whose families were both career politicians attuned to current events as well as career military men. Washington was a hotspot of protest marches; in addition, my adolescent fantasies of sex, drugs, and rock 'n roll, stoked by the 1967 Summer of Love, were increasingly at odds with the discipline of military service.

By 1970 any thought of my application to the Naval Academy

had vanished. I was not alone. As a member of the Alumni Board of the Naval Academy, my father was told there were only 1,005 applications for the 1,000 slots in plebe class. If I wanted in, I would be accepted, no questions asked.

My cocksure response was a resounding "No!" Without thinking, I added, "Dad, these days, only losers go into the military." How stinging and painful was that rebuke. The last of his three sons dismissing out of hand his long-held hope that we might follow in his footsteps. Unfortunately, I never fleshed out my thinking with my father.

As this book reveals, however, I did join the Navy, I was a Navy doctor, and I cared for many of the "greatest generation" during my service at Bethesda Naval Hospital. I hope that path makes up in some small way for my youthful arrogance.

My father died in 2007 and was buried with full military honors at Arlington National Cemetery. His second cousin, the Reverend Kitty Lehman, delivered the eulogy. Her words were kind and thoughtful.

> …We cannot adequately attend to our response to this place and to this occasion without asking its larger meaning. We are here to remember and to pray, amidst the fallen. What can be the meaning of so many lives lost in this war-torn world, in this endless struggle between love and fear? Have we come far enough yet, in our restless pursuit of justice and peace, to confirm their sacrifice? I find it most genuinely helpful to admit that we don't know. Still, the clearest triumph is that we hold life most precious.
>
> Some years ago, I picked up a book by Mark Helprin. Actually, I got it for my husband, himself a naval aviator during the Viet Nam war. The novel was about the Alpini during World War I, or at least I thought it was. As sometimes happens, I ended up reading it before my husband got around to it. The title is *A Soldier of the Great War*. Like all fine literature, it turned out to be about life. World War I was simply the backdrop and the metaphor. The Great War turned out to be life itself. Harper was a sailor and an aviator

of the greatest war of all, the battle for life's ultimate meaning. And so are we all soldiers of that great war.

It is in places like this and at times like this that we pause, to consider in awe and reverence all that is of greatest value to us in this life. We recommit our own lives to serve those goods above all else. We express our intuition that such great goods are enduring in a more transcendent sense than we can even imagine. And in so doing, we define ourselves, and we shape the world for future generations.

For Harper, the overriding goods were family, friends, and community, lived out in service to country, to the wonders of science and technology, and to the ultimate mystery we term "God," the benevolence that we hope and trust comprises the very core and extent of all things, of inner and outer space. That was the sermon Harper preached with his life. It remains for each of us to preach our own. May we do as well. AMEN.

I treasure Kitty Lehman's eulogy. I wish I had said something like it to my father. He had arranged for her to conduct his burial service, "to pipe him over the side." Maybe he knew she would capture the essence of his life better than those closer to him. I wish he could have heard her speak, expressing the gratitude and understanding of the trials he faced.

I wish, too, that my grandmother, Sue Palmer Anderson (née Sue Moore Palmer), had been so comforted at her husband General Anderson's graveside service in 1976. The duty chaplain did his best, but my grandmother was distraught with grief and resentment. At the time, I did not understand her anger, but I do now. What I remember best from the ceremony was the 13-gun cannon salute, the smoke from one round to the next, curling up into the sky and then settling around the white gravestones of his West Point classmates and comrades-in-arms.

On the occasion of the 75th anniversary of the end of World War Two, my family was invited to a liberation ceremony in the Dutch town of Roermond, a town my grandfather's troops liberated in March 1945. I was told: coffee with the mayor, a photo opportunity at a memorial site, a parade of restored military vehicles, and a remembrance service in the town cathedral.

My brother Scott had visited both in 1970 and in 2015, but I had no idea what to expect. I was happy to know the citizens of Roermond were planning a remembrance; their plans dwarfed anything in Parkersburg, Iowa, my grandfather's hometown. Still, would the celebration live up to my expectations? Or would the whole thing be a disappointment? It was one of those times when you simply hope for the best and prepare for the worst.

I managed to obtain an American flag that had flown over the American Cemetery in Normandy and a letter of greetings from Iowa Senator Charles Grassley. I tried to learn a few words of Dutch for the meeting with the mayor of Roermond. At the meeting, my brother Scott was going to show her the Order of the Orange Nassau, Grand Officer, with Swords, the second highest military decoration of the Netherlands government, which General Anderson received in 1947. We wanted the mayor and her staff to know how much we appreciated their efforts.

They planned a Remembrance Service, a memorial service in the city cathedral on Sunday at the end of the three-day celebration. I was asked to say a few words, two minutes max. I jumped at the chance.

The remembrance service was more elaborate and moving than I could have imagined: two hours of testimony, orchestral music, prayer, and quiet reflection. The U.S. Embassy in Amsterdam sent a military representative to Roermond; the U.S. 15th Cavalry Group Commanding Officer, Lieutenant Colonel James C. Cremin, and his wife came from Fort Benning, Georgia. Ominously I worried my remarks to this large, most distinguished audience might disappoint.

When I looked out upon the 2,500 faces, I was overwhelmed. I had given speeches throughout my career, but I had never spoken about my family. My throat got drier and tighter. Someone had kindly placed a glass of water on the podium. An eager sip gagged me, almost going down the wrong pipe. "C'mon, Van Ness; get it together," I said to myself. A brief cough and I was ready to begin. It was time to do right by my grandfather.

First, I expressed my admiration of the people of Roermond and thanked the mayor for including my family. There, I was on firm ground. The words poured out easily. When I brought up my grandfather, it was a different story. My voice cracked. Tears welled up. I was nearly overcome. Stopping for a few beats, I took a deep breath—the advice my wife, Sandy, had given me that morning. She knew how strongly I felt about my grandfather. Regaining my composure, I was able to finish with these heartfelt words, "I believe General Anderson would be firm in his charge to us that it is our responsibility to write the next chapter in the cause of freedom."

The sentiment is not a new one. Many others, such as President Abraham Lincoln or Reverend Dr. Martin Luther King, Jr., have said much the same thing. But the expression was from my heart. For me, the words felt right, for the time and for the place.

To the Front: Grandfathers' Stories in the Cause of Freedom is a retelling and distillation of stories from my life in the cause of freedom. Major General John B. Anderson—aka "Ben," "Andy," "Anderson," and "Granddaddy"—was my mother's father and serves as a starting point of the book. His life forms the backbone of the narrative from his early life in Iowa until his death in Washington in 1976.

Anderson's story is intertwined with many others. My father, grandfather to my three children, had his moments in the sun (Naval Academy graduation followed by sea service and flight training) and times of despair (abject poverty as an orphan, marital difficulties, and failure to achieve flag rank).

I, too, am now a grandfather, to two girls. My stories pale in comparison to those of my father and grandfather and are offered for context; the generations flow one into the next.

To the Front: Grandfathers' Stories in the Cause of Freedom also includes tales of friends, teachers, and classmates' families: Wasyl Soduk, a Ukrainian teenager survived the Bolsheviks, the Nazis, and the harrowing post-World War Two displaced persons camps to raise a proud and accomplished family in the United States; Colonel Armand Hopkins endured four years in Japanese captivity only to be bombed at the end of the war by American planes attacking the *Oryoku Maru*; and Major Aaron Cohn of Columbus, Georgia overcame the anti-Semitism of some in the Old South to lead men of the 3rd Cavalry Group.

While I grew up amongst heroes, I was too young to know it. Now that I am older and the heroes of my childhood are gone, their stories endure. Although it is too late to ask questions of them, it is not too late to learn from their lives. As Reverend Lehman said, "They preached sermons with their lives."

Today, we are facing threats to our republic that call into question so many things. The values of duty, honor, and country our grandfathers relied on to build lives of significance are held in less esteem than the merits of re-writing history, of questioning everything, and of establishing a new world order. I believe the standards of the past and the imperative for change are not incompatible. It takes discipline and study to understand that duty, honor, and country were at times conflated with racism, gender bias, and discrimination. It is our responsibility to tease out the good and separate it from the bad.

The lessons of history are hard things. The merit-based rigors of West Point, the Naval Academy, and the Landon School for Boys molded men capable of enduring the horrors of war and captivity. Today's leaders and institutions can evolve from those solid foundations to be more nuanced and capable.

One last word before we begin. With creative license, and

based on my recollections and research, I have created dialogue between many of the characters and the narrator, me. Historical figures are brought into the narrative with words that I believe are appropriate and consistent with their stories. Some of the stories are controversial, reflecting the attitudes and mores of the times. In presenting them unvarnished, no offense is intended. Rather, I hope the reader will realize how far we have progressed in so short a time.

TO THE FRONT

Grandfathers' Stories in the Cause of Freedom

1

The Back Hall

When I was six years old, in 1959, I began to sleep over at my maternal grandparents' house every third Friday night. I was always excited to go even though at first, I didn't really know my grandparents. To get away from my two older brothers and to be the center of attention was great.

At first, every time I spent the night, my mother reminded me, "Granddaddy has rules. You must be quiet and stay in bed 'til 8."

Later I learned if I had to get up, I could watch TV, but with the volume turned way down low, so as not to disturb Nanny and Granddaddy. After 8 a.m., I'd knock on their bedroom door and peep in. Nanny usually stirred first. "Come in, Michael," she'd say in her high-pitched Southern accent. The bedroom air was a little stale, warm and stuffy like old folks. As I drew closer, Nanny would pat the covers, inviting me to climb up the three steps of the bedside stool. Then, I would snuggle in between the two of them, and Granddaddy would roll away, on to his side, facing the wall. Nanny would pull me close, her arm around my shoulders, my head comfortably on her chest. She'd stroke my hair and ask me how I'd slept.

It was nice in the winter to lie there, warm and cozy, on Nanny's side of the bed. After a bit, she'd ask me to show her what small treasure I'd picked out of the old cigar box, my reward for waiting until 8 to come in. "Oh, that's a nice one," she'd declare, looking at a trinket I held up. "It's a badge from the days we were stationed at Fort Leavenworth, with your mother." Then she'd send me off to get dressed. "I'll make bacon and eggs, OK?" she'd call out.

Easing herself out of bed, smoothing her nightgown, she'd slip on her bathrobe. As I dressed, she'd be in the kitchen, scrambling the eggs and frying up strips of Gwaltney's bacon. She was very particular about her bacon, being from the Deep South and all.

Breakfast was special, a meal planned especially for me. Granddaddy always laid out the silver the night before, including silver coasters under the thick-cut glasses for orange juice. Teacups and saucers, too. I loved the air of formality, a sense of aristocracy. My tea took milk, and two teaspoons of sugar. English muffins, butter, and jam complemented the scrambled eggs and bacon. All set out for me.

After a bit, Granddaddy would come in, a wee bit grumpy. He didn't eat much in the morning, just drank a Coke and maybe some orange juice. His doctors had told him coffee would exacerbate his glaucoma, so he got his caffeine and sugar from an 8-ounce Coke. He'd get "a lift" from it, he said, enough to set off on his list of chores.

Supper was a big deal, too. "Mind your manners at supper," my mother would tell me. "Sit up straight. Take small bites. Chew with your mouth closed. Don't leave the table until everyone has finished. Always help clear the table."

Furthermore, without being asked, I was to go to bed at 8:30 or 9:00 each night. "Nanny and Granddaddy are old, set in their ways," my mother told me. "One other thing, do NOT go into the back hall."

To myself, I'd always say, "Ok, Mom, I got it." I knew better than to say it out loud.

Despite all these rules and advice, my turn every third Friday couldn't come quickly enough. "Yes, Michael, I'll take you over after school," my mother reassured me. "Now get ready or you'll be late for school."

All day long in class, looking at the clock, seconds seemed like minutes, and minutes became hours. Some days I rubbed the worry-stone in my pocket. "Be a good boy. And do NOT go into the back hall," the voice in my head repeated over and over. And then it would be time to go.

As my mother and I walked up a flagstone path lined with English ivy to Nanny and Granddaddy's house, she would say brightly, "Be a good boy. There's Nanny. Off you go."

Then I would trot away, into my grandmother's arms. She'd hug me warmly, pulling me into her ample bosom, all talcum powder and rosewater fresh, saying, "Hey, dear! Come in and get a Coke. They're on the door of the icebox."

As I scampered in, my mother would chat for a moment with her mother, then call out, "Bye, Michael. See you tomorrow."

"Bye, Mama." Then she'd go, and cold Coke in hand, I would join Nanny and Granddaddy on the shaded back porch, a little prince, adored and admired.

One such day, it was late autumn, still warm during the day, but cool at night. Nanny told me how much fun it was having us back in Washington, the family all together again, having visits and talking about school.

Granddaddy didn't say much, just listened to her talk. "In a few days," she said, "we'll button up the house, get ready for the raw weather of Washington in winter. It's the worst kind of cold, gets into my bones."

I nodded, not having a clue. And then Nanny got up, "I'll make us some supper."

Since it was Friday night, it was just the three of us. Other days, Virginia, my grandmother's maid, would prepare them supper. But Virginia had Friday afternoons off, in the tradition of the Deep South. So when I came over, Nanny made supper. In warm weather, it would be tuna fish sandwiches, without the crust, using only the best bread, Pepperidge Farm. While Nanny worked, Granddaddy sat with me, not saying too much, seemingly content to have company. When I finished my Coke, he asked, "You want another?"

I supposed I'd died and gone to heaven. "Another one?" I asked. "You sure?"

Looking up from his paper, he said, "Sure, it's OK. You know where they are."

Heading to the icebox, I thought, "This is great!"

After supper, Granddaddy retired to his favorite spot on the sofa, smoking his pipe and sipping bourbon. Evening was falling now. Nanny returned to the kitchen, cleaning up, and then called out for me to wash my hands, "Use the bathroom in the front hall."

Obediently, I did as she told and then wandered toward the living room. Granddaddy was sitting contentedly. He had poured himself another drink, more intent on it than me. I stood still, not sure where to go or what to do. I was standing near the back hall.

I gazed into the shadows, trying to make out what was back there. A hole in the floor? A crack in the ceiling? Nothing. Moving closer, I saw a big old book, a Bible with a tattered cover, on a side table, and a wind-up clock on the opposite wall, ticking away the seconds. "What's the big deal about the back hall?" I thought, peering in more closely. "I wonder what's down there." Like Adam and the apple, or Odysseus and the Sirens, I was drawn in.

The back hall was dark, shaped like an "L." It bent under the stairs and led to a room, cluttered with books. A heavy, Bakelite telephone squatted on a table next to a rocking chair too large for the space. The air was musty, thick with moisture and mildew. I stood in the doorway, letting my eyes adjust. I could make out a desk, and another table with a pipe rack. I thought, "Where's the light?"

The light switch I'd flipped in the bathroom had a loud snap. Back here, if I flipped a switch, I'd give myself away. So I just stood there. And then I thought, "I'd better get outta here."

Turning to retreat, I saw a photograph in a simple, black wooden

4

frame, hanging on a nail. Ten men or so were in the picture—old Army men, mainly, all in a boat. Some smiling, looking at the camera, some frowning, looking off in the distance. "Who are they?" I wondered. "And why is this picture hanging here, in this back hall, all dank and dreary?" I looked again. My eyes had adjusted to the dark, and I could see it better.

Granddaddy's smile leapt out at me. He was in the picture, one of the Army men. He was smiling straight at the camera. He looked happy, delighted even, surrounded by his friends.

And then my grandmother's voice, "Michael, where are you?"

I replied quietly, "Back here," thinking, "I'm in trouble." Hadn't my mother warned me? I was NOT to go into the back hall!

Coming around the corner, Nanny saw me shrinking into the corner, trying to disappear. Hands on hips, puffed up, feigning anger, momentarily all fuss and feathers, she smiled and brought her face close to mine. My chin was quivering, my eyes filling with tears. "I know I'm not supposed to be back here. I'm sorry."

"No, you shouldn't be back here," she said quietly. "You know why?" I shook my head. "You see this switch? Looks like a light switch, doesn't it? Well, it's not. Don't ever touch it, OK? It'll turn the furnace off. In winter, we'll catch a chill. And it turns off the hot water heater. So we'd have no hot baths. Granddaddy would have to relight the pilot light. He hates to go down into the basement to do that. It's cold, and dark, and smelly, he says. You understand? That's why we don't want you back here." Then she started to lead me away.

I wasn't going to be punished! I wasn't going to get a spanking, or worse, sent home. Recovering, I wiped my nose on my sleeve and pointed, "Why is Granddaddy in this picture?"

She took the picture down. "Let's let him explain," she said, leading me by the hand to the living room. Granddaddy had finished his pipe and his drink. He'd clearly been listening to the whole drama in the back hall. Taking the picture from Nanny, with a twinkle in his eye and bourbon on his breath, he said, "Good thing you didn't flip that switch. Would've been a real headache."

But I wanted to know about the picture. "Granddaddy, why are you in a boat with the Army men?"

He turned to the photograph. "See, that's me, next to my friend, General Simpson. It was 1945, not quite twenty years ago." Then he named the others, Field Marshal Montgomery and Prime Minister Winston Churchill—just names to me, but clearly his friends. "Dwight Eisenhower was there, too. We had breakfast together. But then he left."

Up to that point, Nanny had been smiling. At the mention of Eisenhower's name, she stiffened. "Bootlicker. Eisenhower was just a bootlicker."

Granddaddy looked at her sympathetically. "Honey, we've been over this a thousand times. No need to fret yourself," he added firmly.

Then to me, he said. "Ike warned me Churchill would want to cross the Rhine, to see the fighting up close and personal. I tried to say no, but Monty overruled me. When this picture was published on the front page of every major newspaper in the Western world, Ike blamed me, thought I had defied him, his old friend. And sometime later, he got even. Took his revenge cold—not then, but later."

Nanny now regretted that she had started this. She'd seen it too many times before. After a drink or two, Granddaddy would go on and on. It was sad that Granddaddy and Eisenhower, two old West Point war horses who had fought side by side in common cause against despots and dictators, were now estranged. A kind word or gesture from either of them might have eased the memories of the horrors they endured. Instead, the events of the past haunted Granddaddy and alcohol heightened the pain. Nanny knew I was too young to understand. Maybe someday, but not today. Time to nip this conversation in the bud. She said, "It's bath time. I'll draw you a hot tub, Michael. Come with me."

Then it was time for bed. As she tucked me in, she said, "He'll tell you some war stories tomorrow."

And he did, on that day and on many others over the years. My grandfather's life, from Iowa farm to command of the largest combat corps in World War Two, is a powerful story of service to country, dedication to the Army, and devotion to the ideals of West Point. It is heartbreaking at times, funny at others. Now that I am the age he was when he first started sharing his stories with me, I continue to learn more of the true significance of his service, and I am ready to share some of his war stories, and mine. Everything I learned from him has helped shape the man I am today. I recognize his struggles with the demons of loneliness, rejection in love, and disappointment at the end of his long Army career. By the time I knew him, he was nearly 70 years old and his broad smile shone only rarely on me. I am so grateful to see his grin in Army newsreel footage I discovered during my research.

2

Landon School Summer Day Camp and Landon School for Boys

My first summer in Washington, D.C., 1959, was a time of transition, from the West Coast to the East, from a military base to the suburbs, and from the top of the heap to the bottom. As dependents of a commander in the Navy, on a base inhabited by young officers and enlisted men, among aviators and rocket scientists, my family had been special, known to most of the military police, doctors and nurses, and civilian contractors. Not so in Washington.

We were now fish out of water, just another military family in a city at the center of the world stage. My father was among the first to join the ever-growing staff at the National Aeronautics and Space Administration (NASA) charged with countering the threat posed by the Russians and their *Sputnik* satellite. The Russian achievement challenged the notion that the United States was the most advanced nation in the world. It challenged our prestige as the best hope for the future of developing nations. When Yuri Gagarin became the first man to orbit the Earth, the demand for action increased. The United States was no longer the leader in space exploration. We had lost that position. We were no longer the best. The whispers were starting, "If the Russians could put a satellite in space, what else could they do? Maybe communism was the way to go, not capitalism."

Sputnik was also seen as a possible means by which the Russians could hold the high ground and thus have an advantage over us in any military conflict. Thus, President Eisenhower had mobilized the military to get in the fight, especially after the televised spectacular explosion of the Vanguard TV-3 rocket on the launchpad at Cape Canaveral. Too many high-profile launch attempts were ending in failure. Too many people began to doubt our leadership. Too many laughed when Soviet Premier Nikita Khrushchev called our first successful satellite in orbit "a little grapefruit."

My father, Harper Elliot Van Ness, Jr., was trained in rocket science. At Point Mugu, California, he had been a key member of the team that the Navy had put together to develop a series of submarine-launched missiles, the Regulus missile system. In fact, missiles launched from surfaced submarines had been flown remotely and landed successfully at Edwards Air Force Base, about 50 miles in from the coast. And so, we left Point Mugu, the California beaches, and a comfortable life at the top of the pack for the world of politics and power that is Washington, D.C.

I did not know the reasons for our move; I simply knew that I was not going back to Ocean View Elementary School for second grade. I did not really have any bosom buddies I was going to miss, but I cried nonetheless, likely in sympathy with my mother, who had a true friend in Mrs. Nancy Prothro, the wife of a hard-charging F4 Phantom pilot. My mother, Sue Moore (Anderson) Van Ness, was very sad to be leaving Nancy, a Louisiana beauty with seven children who was happy-go-lucky in a way my mother never seemed to be. When Mrs. Prothro took up hula dancing, she included my mother in the group lesson at her house. Of course, I was dragged along, only to sit watching these Navy wives' hips swaying to and fro:

Oh, we're going to a hukilau
A huki huki huki huki hukilau
Everybody loves the hukilau
Where the laulau is the kaukau at the luau
We throw our nets out into the sea
And all the amaama come a swimming to me
Oh, we're going to a hukilau
A huki huki huki hukilau

"The Hukilau Song" had come about one night in 1948, when composer Jack Owens was suffering aches and pains from an evening of hukilau dancing. Unable to sleep, and inspired by the rhythms and movement of the dancers, he wrote the lyrics. The song became an instant hit.

I just sat on the floor, listening to the Hawaiian music and watching the ladies dance. They smiled at me from time to time; I think they were amused by the intensity of a five-year-old's attention.

That first summer in Washington, my mother did not know what to do with her three boys. My father was already deep into his work at NASA, dedicating all his time and effort to stopping the Russians. He could not be distracted by events at home; it was the job of the good military wife to "keep the home fires burning."

Even though Washington had been my mother's home in the 1940s, fifteen years later, it was a very different city. Any connections she'd had with friends and clubs were broken. For a Southern debutante like my mother, the changes were all very discombobulating. She thought she needed to figure a way forward by herself. She wanted to show both my father and Nanny that she could keep her boys busy and that she could figure out arrangements for school starting in the fall. First, she tried a private pool in our neighborhood. It was expensive to join, but courtesy of a neighbor who was a member, we had a free trial swim one afternoon. That afternoon was a good one, but too short-lived, and not a remedy for our long-term needs. And it did not fit our family budget.

Then my mother learned about a possible opening for me for the coming school year at the Beauvoir School, the elementary school of the National Cathedral and "Old Washington" society. Whereas politicians and their minions came and went, if you were Old Washington, you had it made. I think my mother considered us Old Washington. Let's do Beauvoir!

If I got in, she thought that the school's sibling policy would soon let her other boys in. Then, Beauvoir would lead directly to the prestigious St. Albans School for boys—if not for the upcoming year, then the year after. But I still had to take the entrance exam to make my acceptance official.

I suppose my mother thought I would be a shoo-in; what mother doesn't think her child hung the moon? Since my mother had attended the prestigious Mount Vernon Academy for Girls in the years leading up to Pearl Harbor, wouldn't that tip the scales in my favor? Since she was married in the Washington National Cathedral in 1947, she was Old Washington. And let's not forget, her parents were founding members of the Army Navy Country Club in Arlington. Her father was a retired general. And her parents were once members of Chevy Chase Country Club. "Didn't we old-schoolers take care of one another?" she must have thought.

The day of the entrance exam for Beauvoir was hot and humid, a typical July day in Washington. Two of us took the test for a single spot that had opened unexpectedly that month. An hour or so of pictures and puzzles that seemed to be no big deal, easy, or so I thought. Turns out, I came in second.

The admission's director spoke kindly and was very solicitous of my mother's feelings about my failing to make the grade. She offered an alternative: I could enroll in Beauvoir's summer camp program. She showed us the playground, a confined space behind the school, up above the expansive playing fields of St. Albans School for boys. If I liked the camp and became known to Beauvoir, I could apply for next year.

Her suggestions made my mother angry. No matter how the director put it, I was "out." The whole experience showed that our family lacked the deep roots, the longstanding ties through marriage and property, or the clout to simply call someone and gain entrance to the very best schools. That's what really hurt.

My mother swallowed her pride, smiled, and thanked her for her time. Not one to make waves, she led me to the car without a word. As we climbed in, it was stiflingly hot, having sat all morning. As we pulled away, down came the windows, to catch a breeze and blow away the cigarette smoke. My mother loved to smoke and my father hated her habit. She rationalized it by saying she wanted to keep her weight down; my father thought she should do so with exercise, not smoking. On occasions like this one in the car, she would light up, knowing he wasn't around and I couldn't care less. (In fact, I sort of liked the smell of cigarette smoke. When I came into the bath-room shared by all, the lingering smoke previewed one of my great-est childhood fetishes: I would take aim at the cigarette butt in the toilet and fire away, like a fighter pilot hosing down an enemy plane. If my aim were true, I could move the butt up against the porcelain wall and pound it until it came apart, exploded by my steady fire. What fun that was!)

I was happy to get away from Beauvoir that day, to get out of the tie cinched around my neck and the three-button blazer that was making me sweat so. I'd never had to wear ties or jackets at Point Mugu, where flip-flops, shorts, and bare chests were standard summer garb. My mother was glad to get away, too; humiliated, her pride stung, and she needed a plan—and another cigarette.

When my grandmother heard the news, she wasn't fazed. She had already consulted the Army network of retired officers' wives who knew each other from tours of duty all over the United States. Although Nanny had gone along with the idea of my application to Beauvoir, she thought the Landon School would be a better idea, not just for me but for all of her grandsons. They had a summer program for boys aged 6 through 12. The program was held on the school campus, an old tobacco farm on several hundred country acres, not just a few groomed fields hemmed in by city streets like Beauvoir and St. Albans. Landon's program could take all three of us—Scott, John, and me—in their second summer session beginning in a few days. It would get us out of the house, taking a lot of pressure off our mother to keep us occupied. And Nanny knew the camp led to acceptance of many campers at Landon School. That's what she really wanted.

Before the week was out, my mother had us three boys enrolled in the Landon School Summer Day Camp. The first day, she took us hand in hand down a long grassy bank toward the swimming pool to meet the swimming instructor, Mr. Tom Dixon.

My swimming endeavors had gotten off to a scary start earlier that summer during our cross-country drive from California. We had stopped at a lake near Lake Tahoe one day and my brothers and I jumped in the water. Not wishing to be left alone as my brothers swam away from shore, I bobbed along behind them, going deeper and deeper into the water. With each step, I would hold my breath, sink down until I could touch the bottom, then push off. Breaking the surface, I would then catch my breath. Farther and farther from shore I went, gasping for air, stepping forward, sinking, and then bouncing up again. And then I couldn't feel the bottom.

Sitting on the lake shore, my mother saw me go under, not knowing if I was going to bounce off the bottom again or not. She saw my brothers swimming away from me, not paying any attention.

"Michael's drowning!" she thought. No one else saw what was happening. Springing up, throwing cigarette and drink aside, towels flying from her shoulders and waist, she charged into the water. Bathers scattered out of her way. She was a wild woman on a life-saving mission. She grabbed me around the waist and dragged me to shore. "What were you thinking?!" she screamed at me.

"I dunno." I replied, not knowing how close I was to disaster.

"Just following Scott and John."

"That's it!" she said. "That's never going to happen again." Turning to my father, she added, "When we get to Washington, Michael is going to learn to swim."

There we were, with the new swim instructor, Tom Dixon. She wanted to be sure he knew the story of my near drowning. Tom Dixon took me under his wing. He doted on me until he could see that I was not going to drown, and then challenged me to get better and to swim the length of the pool underwater. When I mastered the Australian crawl, he added the breaststroke and sidestroke to my repertoire. With his encouragement, I learned to swim that summer, well enough to allay my mother's fears, and even to win a race or two. It was a great summer experience. When honors were passed out, I wore my ribbons and medals as proudly as any military veteran. And I've never feared going swimming; for that, I am forever grateful to Tom Dixon.

Tom Dixon lived at the Landon School, a confirmed bachelor. He loved his Beethoven and his books. He was my English teacher in later years at Landon, where Shakespeare was his area of particular interest and expertise. He did his best to imbue his young charges with the beauty of the language and the depth of Shakespeare's insights into human nature. When I boarded there my senior year, he was the senior dormitory house master. By that time, we had known each other for over ten years. Because of that history, despite the natural tendency for adolescent boys to push the limits, Mr. Dixon gave me privileges to come and go as I pleased, to go out to the theater on school nights, and to stay up later than "lights out" to read or catch up on my work.

Tom Dixon was also a military veteran, a rifleman in the 104th Infantry Division commanded by the notorious General Terry de la Mesa Allen. After Patton, Terry Allen was the most colorful of the combat commanders in World War Two. In North Africa, in 1942–43, Allen led the First Infantry Division to a series of victories over the vaunted Wehrmacht, driving the Germans out of Tunisia. Allen's men adored him for his swashbuckling ways and for the lax discipline enforced during times of rest and relaxation. Eisenhower and Bradley were less amused and felt that Allen's troops lacked proper control, with conduct unbecoming of United States soldiers.

General Allen was too good a leader to be left out of the war, too good a fighter not to get another shot at the Germans. He was given command of the new 104th Infantry Division, the "Timberwolves," and applied the real-life lessons learned in North Africa to create a first-class assault division. Omar Bradley, who relieved Allen two years before, ranked the Timberwolves among the best, as good as the veteran First and Ninth Infantry Divisions.

Tom Dixon told us boys that, "Yes, I fired my rifle in combat. But I never aimed it, and never killed anyone, that I know of." When he started a biography of General Allen, Mr. Dixon interviewed Granddaddy about him. Unfortunately, Mr. Dixon never completed the work, nor published any insights from his interview with my grandfather.

After the summer camp session, my brothers and I attended a year of public school at Radnor Elementary. Like all the schools in the area in 1960, Radnor was integrated, a new development for the D.C. area, as a consequence of the recent end of the so-called "separate, but equal" segregation laws overruled by the Supreme Court's *Brown v. Board of Education* decision in 1954. Class size was about 30, but in those days, the teachers had complete control, supported by post-war parents eager for their baby-boomer children to conform and learn. There was never any racial tension of which I was aware, but my grandmother Nanny was determined that her grandsons would remain in public school as short a time as possible.

Nanny, nee Sue Moore Palmer, was a product of the Old South. Raised in Columbus, Georgia, on the border with Alabama on the banks of the Chattahoochee River, her family revered General Robert E. Lee and the "Lost Cause." In fact, in the foyer hung portraits of Lee and George Washington. She spoke lovingly of her childhood maid, Sarah, who was Black. Nanny often said of Sarah, "She was a member of the family. She made the best fried chicken. Always in lard. I loved her as much, if not more than my mother." Underneath these sentiments lurked Nanny's undeniably racist political views. It was as though the Civil War and Reconstruction had ended just yesterday, and her Southern Democrat views barely concealed her resentment of the carpetbagger Republicans she used to justify her prejudices. Granddaddy's accounts of heroic service by soldiers of all races failed to deter her. Occasional outbursts over the Civil Rights Movement and busing made me cringe, especially when she resorted to the coarse vocabulary of her youth.

There were no Black students at the Landon School for Boys in Bethesda, Maryland. The school was segregated until 1963, with the admission of a single, brave young man named Alexander Aikens, who ignored whispered racial epithets with quiet dignity and grace, preferring to study than to respond to bigotry and hate. I remember seeing Alex but don't recall speaking to him. The racial divide was a real one, but in truth, fifth graders like me had few reasons to approach any third-former. His quiet determination gained him the respect and admiration of faculty and classmates. He graduated in 1967 and attended Brandeis University.

Alex entered Landon in the autumn of 1963 as a third-former, four years after my brothers and I took the entrance exam. My brothers started in 1960, while I waited to start third grade in 1961. My grandparents provided the financial wherewithal to cover tuition, athletic fees, and sport store charges. Whereas my mother accepted their help without protest, my father was embarrassed by the need for assistance. He had always banked at least 10% of his Navy salary into a "rainy-day fund" and knew Landon's fees would ruin the family budget.

My grandparents' reaction to Alex's attendance and Banfield's decision to desegregate were muted, most likely given the paltry number of Black students during my brothers' and my time at Landon. My parents were unfazed by the change, going along with the times and more concerned that their boys excel in both their studies and athletics.

And so, in September 1961, I began my ten years at the Landon School for Boys. I wore a coat and tie every day, beginning in third grade. After third grade, all the teachers, except for the art teachers, were men, and many were veterans; we called them "masters," in the English schoolboy tradition. When adults entered a room, all the students stopped whatever they were doing and stood up. It was always, "Yes, sir" or "No, sir," "Yes, ma'am," or "No, ma'am." A student initiative led to the adoption of an honor code, similar to those of the Naval Academy and West Point: "We neither lie, cheat, nor steal, nor tolerate those who do." The honor code was proposed by Landon's Student Council in the early 1960s. The faculty and staff were skeptical. Education of the student body and many of their parents was needed. In practical terms, the Honor Code meant tests were not proctored and lies were not tolerated. If one were to cheat and get caught, expulsion was expected. If one were to observe someone cheating, it was expected that the observer would report the cheater. Failure to report a cheater was considered as pernicious a fault as cheating itself. Stealing was never acceptable. Once we learned what was expected of us, the Honor Code was voted upon by the Upper School students and adopted unanimously.

In the 1960s, Washington D.C. was still a Southern town in taste and custom. Segregation was a reality, with tenements and slums in the southeastern, majority-Black parts of the district, distinct from the leafy, majority-white suburbs northwest of Rock Creek Park. Despite the changes in the law, class and racial distinctions were understood and widely accepted or at least unchallenged. African American women in starched maids' uniforms walked from the bus at Westmoreland Circle into neighborhoods like Albemarle Hills, where my grandparents lived. Whether in summer heat and winter snow, their maid, Virginia, quietly made her way up and down the steep hills, in walking shoes she would replace with dress shoes before entering the house. With the passage of the 1964 Civil Rights Act, the high-minded rhetoric of the *Washington Post* and the political class contrasted with the reality of those who liked the old ways, like my grandparents. Change was in the air, but resistance to change was fierce.

The school's founder, Paul Landon Banfield, was more interested in character development and athletics than politics or the Civil Rights Movement. He envisioned his school, founded in 1929, as a cross between an American military academy and an English private school, or a civilian military academy. He was guided by the writings of men like Colonel George Taylor, who wrote, "Personality is the most important element in the conduct of war. It may be said that it affects almost everything." Taylor also wrote, "In a landing operation, there are two classes of men that are found on the beach: those who are already dead, and those that are about to die." The harsh reality of his statement was proven at Omaha Beach. Banfield was determined to develop his boys into strong and capable men, future leaders of men, guided by the school motto, *Virtue et non vis*, translated roughly as "right over might." Mr. Banfield articulated the school philosophy:

> The philosophy of the Landon School is to develop a growth of individual responsibility, leading to a group consciousness and an awareness of the rewards and penalties of selfish and unselfish participation. In this way we feel that we can prepare our boys for changes within our social structure, be they abrupt or gradual, with a minimum of maladjustment, and that they will be useful and happy in any order of society.
>
> To this end the school as a whole encourages and recognizes growth of individual responsibility in staff and pupils, rewarding unselfish response with more responsibility and penalizing, usually by public opinion, selfish usurpation. The masters have freedom to develop

18

their boys as well as themselves. To have the things they want for themselves and their school, the boys must conceive and develop their wants. We have had the experience many times of seeing the thoughtfulness of one or more boys overcome the thoughtlessness of others, thus a responsible leader always seems to appear.

We believe that with the union of individual responsibility, and the democratic processes, is born a disciplined citizenship.

Grades were posted by name for all to see. Honor rolls and grade point averages determined class ranks. Athletics were compulsory, three seasons a year. Like the boys at Eton College (a university preparatory school for the ruling families in England), every boy played sports, regardless of innate ability, in all types of weather. Many of the masters also coached, the thought being that character development occurred both in and out of the classroom.

It was a non-sectarian school, but morning assembly started with the Lord's Prayer. We sang fervently Episcopalian and Presbyterian hymns and said the traditional Catholic meal-time prayer before lunch:

Bless us O Lord
and these thy gifts
which we are about to receive
from thy bounty,
through Christ our Lord.
Amen.

We didn't recite the entirety:

May the souls of the faithful departed
through the mercy of God
rest in Peace.
In the name of the Father, Son, and Holy Ghost,
Amen.

Although not raised in the Catholic tradition, I like the entire prayer. Maybe it is a deep-seated yearning for a more religious view of the world, something on which I am still working. As C.S. Lewis notes in *Surprised by Joy,* it is important that we know Heaven exists, more important than that any of us should reach it. In the meantime, I continue to enjoy the entire traditional blessing and prayer and say it often. In the Anglican tradition, I cross myself at the end of the prayer.

Befitting an institution dedicated to disciplined citizenship, military service was common. With the outbreak of World War Two, sixteen masters volunteered. For example, despite four years of college and experience as a teacher, Mr. John H.F. Mayo heeded the call to arms and enlisted as a private in early 1942. Unfortunately, he was killed on October 4, 1943. He was not alone in that sacrifice. More than 300 Landon graduates and non-graduates joined the Armed Forces, and twelve of these "Landon men" were killed, most in the prime of life.

Among them was John R. Amussen of the Class of 1940. Amussen was a top student and athlete. During a game against the Sidwell Friends School, he committed an error that resulted in a base-clearing triple. Mr. Banfield was coaching the team and berated Amussen in front of the entire team, saying "I suppose you learned that at the A&W when you missed practice yesterday afternoon." Benching Amussen for the remainder of the game added insult to injury. Compounding Amussen's dismay was the truism of the day that one bad mark follows you wherever you go till you die.

Upon learning that Amussen had missed practice to care for his sick mother and to perform necessary household chores, Mr. Banfield offered a public apology. In front of the Upper School at lunch, Mr. Banfield said, "As long as I had committed an error in publicly berating Amussen, I also owe him a public apology. He missed practice not to go to the A&W, but to care for his mother. I hope he will be gracious enough to forgive me my temper tantrum."

Amussen shook Mr. Banfield's proffered hand, and said, "Sir, anything you want from me from now on, you can have."

Amussen joined the Navy out of Haverford College, flew combat missions in the Pacific theater, and was awarded the Navy Cross and two Distinguished Flying Crosses. He was killed in a training accident at the end of the war, never fulfilling his dream to return to teach at the Landon School.

Death-defying service was common in that generation. Some of the parents of my class of 1971 were Landon alumni who were also in combat. For instance, David J. Dunigan, Jr., the father of my classmate Bruce Dunigan, flew 35 missions as a co-pilot and pilot of a B-24 bomber of the 467th Bomb Group, stationed at Rackheath, England. Dunigan was decorated with the Distinguished Flying Cross, the second-highest medal for valor after the Congressional Medal of Honor. Another classmate's father, Richard S.T. Marsh, left Yale College to fly for the U.S. Marine Corps, VMF 321, during the darkest days of the war in the Pacific. My classmate Tom Fuller's father had a difficult time in the service. Transferred unexpectedly out of the newly arrived 69th Infantry Division, he was thrown into combat with the 83rd Infantry Division. Many replacement infantry men did not survive; they were inexperienced and without friends in their new units. In fact, veterans often ignored or neglected the new arrivals. Against the odds, Fuller survived the Battle of the Bulge and came home alive.

My wrestling team comrade, Tom Hamilton aka "Hawk," told a story about his father, Lt. Lloyd Hamilton, a World War Two B-24 pilot. Lt. Hamilton flew on the some of the most hazardous missions of the war—the attacks on the Ploesti oil refineries, where flak was so thick it was said you could walk on it. When Tom suggested to his father to watch the movie *Twelve O'Clock High*, his father replied, "Thanks, son, but I caught it the first time."

My second wife Sandra's father, Charles Carlton Elsbree, despite a history of mastoid surgery that usually precluded military service, started out World War Two with hopes of becoming a pilot. In the early stages of flight training, Elsbree rolled his plane, causing his instructor pilot to lose consciousness. The result? No more stick time for Elsbree. Despite washing out of pilot training, he remained in the U.S. 5th Air Force as a B-24 crew member. On his way to invade Japan, the war ended. Elsbree was sent to Korea to search for missing American prisoners of war. His duty completed, he started home. On his way, he visited Hiroshima to see for himself the widespread effects of the atomic bomb blast on the city and its populace.

On Veteran's Day, my Landon classmate John May reminded all of us of its origins, to remember the fallen from World War One on the day the guns went silent on the Western Front. His father was born in Cornwall, England and joined the Royal Air Force in 1941. Trained in a Tiger Moth, May flew the Avro Anson, Wellington, B-24 Liberator, and the B-25 Mitchell. A recipient of the Distinguished Flying Cross, May married Gwen Jones, whom he had met on Christmas Eve in 1943 while on leave in Washington, D.C.

Many of Landon's masters were retired military officers. It was only later, when we were enrolled in other institutions with different traditions, that the nature of our formative years came into focus. In fact, it was often only by comparing recollections that we understood how it had happened. For example, unknown to me in 1968, my Fourth Form French teacher, Colonel Armand Hopkins, was a West Point graduate. He was firm and fair, inspired in his quiet determination that we would master the past perfect tenses of French verbs, both regular and irregular. I remember his saying, "The best way to prepare for the final examination in French is to prepare every day."

This was a great lesson for any young man, whatever his future, but an essential lesson for a future soldier, a doctor, or a teacher. Hopkins would say, "If a soldier has not prepared well for combat, he has no chance to survive. Just like a doctor before an operation, or a teacher before a class, the soldier must be ready beforehand. And so boys, let's start now today to prepare for the final exam."

I never before experienced such perfect courtesy in a Landon master. That observation is not meant to suggest softness. It does not. Instead Hopkins expected us to live up to his standards and would not accept less than our best.

On occasion, he could be severe and take a more direct approach.

"I learned Japanese at the point of a bayonet!" he would exclaim excitedly. Then he'd slap a ruler on the desktop and count in Japanese, "*Ichi, ni, san, shi.*"

He was not taunting us; he was leading us. Like a judge in a court of law, weighty and measured he added calmly, "If I could learn Japanese at the point of a bayonet, you can learn French!"

With the ruler as a baton, he'd continue: "Repeat after me: *je suis, tu es, il est, nous sommes, vous êtes, ils sont.* OK? Let's do it again."

On occasion, we tried to turn the tables. Boys being boys, noticing most everything wrong with one another and our masters, we whispered among ourselves, "Colonel Hopkins doesn't have a right ring finger." We dared not ask him about it. We knew he would not answer, and likely would think unkindly of anyone who asked him such a personal question. By his demeanor, we knew he would not be distracted from the daily French lesson it was his duty to teach. He was not there to tell us his personal story. I wish he had. It might have blunted the burning desire to creep up the hierarchy that dominated the academic and athletic life of Landon. It might have increased our understanding of the risks of a life in the desert of burning ambition.

I discovered the true story recently. In 1984 Colonel Hopkins wrote *Reminiscences of Armand Hopkins*, principally for his children and grandchildren, and submitted the work to the library at West Point. In it, we can find the details of his wartime experiences.

Hopkins graduated West Point in 1925. Like my grandfather, he had a variety of assignments in the inter-war years. Arriving in the Philippines in October 1941, he was assigned as Executive Officer of Fort Hughes on Caballo Island, near the entrance to Manila harbor, a rocky island, part of the rim of the sunken volcano that also includes Corregidor. With the attack on Pearl Harbor in Hawaii and Clark Air Base in the Philippines, war was declared in December 1941. The United States was ill-prepared and steadily gave ground to the Japanese attack in the Philippines, ultimately pushed into the Bataan peninsula. The Americans and our Filipino allies resisted until food and ammunition were exhausted. Falling back to Corregidor, MacArthur planned a last stand on the island. President Roosevelt ordered him to return to Australia to organize an army to return to the Philippines in the future. General Wainwright was left in command and ordered to surrender on the May 5, 1942.

The humiliation of the Americans and Filipinos was complete. All American resistance was ordered to cease. Despite initial reassurances by the Japanese that humane treatment would be given to all who surrendered, Colonel Hopkins and his fellow West Point officers suspected differently. The Japanese were disdainful of soldiers who surrendered; they regarded it a betrayal of the Bushido Code by their soldiers, and a mark of cowardice in their enemies, especially among officers. Regardless, following his orders, Hopkins personally supervised the raising of the white flag of surrender to the Japanese swarming over Caballo, and the ordering of the funereal-like burning of the American flag, keeping it out of Japanese hands.

Colonel Hopkins endured more than three years of captivity, first at the infamous Cabanatuan Camp #1 in the Philippines, later in Japan and Korea. Japan did not abide by the Geneva Convention of 1929, dictating the rules by which prisoners were to be treated. At first, when the Japanese were reeling off victory after victory, Red Cross packages arrived and food was present in sufficient quantity to stave off the worst effects of malnutrition. After the United States victory at Midway, however, treatment by the Japanese got worse, much worse.

Hopkins kept up his morale by seeking out West Point graduates among the hundreds of prisoners. The men shared news and stories; when they heard good news, they shared it. For example, when a prisoner driver returned from transporting various supplies for the Japanese, he had a story to tell. He had encountered an elderly Filipino padre and asked, "Father, now that the Japanese are in charge, how do Filipinos feel about the Americans?" His answer: "My son, I wish I had as much faith in the Almighty as I do in the Americans."

Hopkins' most harrowing experiences began in December 1944. The Japanese began a series of prisoner evacuations to Japan. Herded aboard the *Oryoku Maru*, the men suffered an attack by U.S. Navy pilots who were unaware that their countrymen were trapped below deck. Later, Hopkins was transferred to a filthy coal ship, the *Enoura Maru*, for transit north to Kyushu. With inadequate food and water, no sanitary facilities, and stifling heat in the unventilated compartments caked with coal dust, prisoners had precious little to hold onto, except each other. Among the 1,900 prisoners who had embarked in Manila, only 500 or so survived starvation and the elements as they traveled the gauntlet from the tropical heat of the Philippines to the snow and cold of northern Japan. Twenty to forty bodies a day were hoisted out of the filth and fetid air of the merchant ship's hold, to be tossed into the sea without ceremony. The second ship was attacked again, with dozens killed, including Leo Paquet, West Point Class of 1919, who succumbed to a shrapnel wound to the chest. Even after disembarking in northern Japan, more prisoners died. In the end, only 300 of the 1,900 remained.

Hopkins stayed alive, desperate at time for food and water. The Japanese used a few of the younger prisoners to distribute the daily ration of one rice ball and water twice a day. It was not enough, especially since diarrhea increased dehydration. Hopkins' good friend Jean Harper lost his fight with malaria, in large part because he could not stay hydrated. As Harper lay dying, Hopkins promised Harper that he would return Harper's West Point ring to his wife.

About this time, Hopkins ate snow from the deck of the ship during a rare "exercise" period. It offered no relief. Down to 95 pounds, he was desperate. Stupor and delirium were overtaking him. He had to do something, or he would die. Hopkins knew one of the younger lads well enough to make a deal: his West Point ring for two canteens of water. He shared the water with his companion, Alex, a fellow West Pointer, Class of 1925, and a survivor of the Bataan Death March. Alex had commanded an infantry regiment during the desperate fight against the Japanese in early 1942 and was awarded the Distinguished Service Cross for unusual gallantry and bravery in combat with the enemy.

During the trip north, in the unheated hold of the prison ship, the two of them slept spoon fashion, knees bent with belly and chest curved around one another, shifting every half hour or so, bones sore from lying against the hard steel decks. Hopkins added, "This way of sleeping, two men close together, their arms around each other, must evoke in the reader a suspicion of homosexuality. It certainly aroused in us no such thoughts or feelings. Our whole concern was to get warm and stay alive."

The two men survived and were ultimately reunited with their families. Alex was crippled with chronic heart disease and died in 1963, his heart disease aggravated by beriberi contracted in captivity. In his memoir, Hopkins said, "Alex was the most patient and loyal friend any man could hope for."

Hopkins made it a point to go to class reunions to see the men with whom he had endured so much. He also made sure to give Cecile Harper, wife of classmate Jean Harper, the West Point ring Hopkins removed from Jean Harper's body aboard the *Enoura Maru*.

By the way, I learned that Colonel Hopkins had lost his finger in an accident, not by an enemy soldier. He was able to get a replacement for his own West Point ring and wore it proudly until the day he died. He wore it next to his wedding ring, which he had hidden on his body throughout the war, kept on a string tied around his waist and out of sight for more than three years.

At Landon School Summer Day Camp and the Landon School, heroes surrounded us; we just did not know it at the time. I will present more of the masters' stories and pictures, and the stories of the parents and grandparents of my classmates, which continue to inform the lives of their descendants and their friends, including me. They were men and women whose lives were full of hopes and dreams, accomplishments and disappointments, and bitterness and joy. It is strangely comforting to know we all are all "soldiers of the great war of life," a phrase from author Mark Helprin's book *A Soldier of the Great War.*

3

Grace

Before supper during my visits, Granddaddy and I would loiter in the living room, waiting for Nanny to finish the preparations. He sat on the old heavy Victorian sofa with its carved wood headrest and green velvet upholstery, worn and comfortable. Like many of the furnishings inherited from Southern ancestors, it was "old money," not flashy and certainly not new.

His spot on the sofa had a depression from years of use, the springs sagging to accommodate his bottom. To his side, there was an end table with his can of tobacco leaf, his pipe, and an ashtray. A green-glazed clay pipe rest, my arts and crafts project from summer camp, held his pipe, the bowl well-charred from years of use, always at the ready for the next bowl of leaf. I'd play on the floor with my Army men, metal tanks and trucks, and airplane models scattered about the old Oriental carpets.

When Granddaddy was teaching me to play poker, he'd call me over, saying, "Michael, why don't you get the chips. Let's play a few hands before supper." Sitting on the green-velvet ottoman, I would deal the cards onto the butler's table situated in front of the sofa. To keep cards from going over the edge, Granddaddy would flip up the hinged side, a handy feature of the antique. Over time, I got better at shuffling and bridging the cards, a fancy trick to keep them from being bent too much. When I forgot to place the shuffled deck for Granddaddy to cut, he'd pull out an imaginary pistol and "plug" me.

Granddaddy usually had me deal, so I had the privilege of calling the game: five-card stud, draw poker, or my favorite, seven-toed Pete. I would also call the "wild" cards—not just deuces, but deuces and treys, or deuces and one-eyed jacks. These hands would end up in wild twists and turns, four-of-a-kind or straight flushes. In seven-toed Pete, betting went back and forth, one of us starting the bet, the other either standing or raising the stakes. When Granddaddy raised his bet, I'd try to make out whether it was a bluff or not. On the other hand, rarely could I hide my excitement at a good hand. If I made an outrageous bet, or tried too hard to bluff, he would just chuckle and "go light" to cover his bet and call the hand.

On occasion, I would win a big pot. After raking it in, leaving one for the ante of the next hand, I'd stack the chips carefully in front of me. I loved to see them organized red in one stack, blue in another. It was a way to gloat, without crowing out loud.

"Michael, be careful. Smirking like that, stacking your chips in front of your opponent is unwise," he'd say. "It's an invitation to 'frontier justice.'"

To be sure that we kept it all straight, he'd pull out *According to Hoyle*. He'd show me where three-of-a-kind beat two pair, or four-of-a-kind beat a full house. Back and forth we'd play, chips gathered in and piled up, or lost on a dare. Granddaddy taught me the rules and rituals, stressing that in an honorable game there were no hard feelings; the cards told the tale. After an hour or so, and two Cokes and a bathroom break, it would be dinnertime. Granddaddy finished his drink and added in an English accent, "Here's how." He learned the expression at Ypres with the British troops in the line. When they finished a drink, usually gin with bitters, they'd announce, "Here's how," to encourage others to hurry up and finish so they could pour another.

At dinner, Granddaddy ate quietly. Nanny wanted to know about our game, who won, who lost. I told about a few hands, Granddaddy listening to the story. When conversation lagged as it often does when children visit their grandparents, I sometimes asked, "Nanny, what was it like in olden days?"

Ah, what was it like in the olden days? A difficult question to answer. What are the olden days? The days before electric lights or automobiles? Or would Nanny tell me about the "real" olden days about the family in the South before the Civil War? Nanny would oblige me with a story or two about Columbus, Georgia after the war or Knoxville, Tennessee, her family's home before what she termed "the war of Northern aggression." Then, with images of Confederate soldiers in my head, I'd head off to bed.

On subsequent visits, while playing poker, I'd ask Granddaddy about the olden days, too. I really wanted his war stories. And so, little by little, often after another bourbon before dinner, Granddaddy found himself back in time.

"Michael," he told me, "I graduated West Point in 1914. My class was small, only a hundred or so cadets. We were together for more than four years, our formative years. We knew each other well, intimately even, as only fraternity brothers in a world removed from day-to-day life can. You see, once you committed to the Army, little else mattered. No outside visitors except on rare occasions like dances or football games."

"Did you like it?" I wondered aloud.

"Not at first. I was terribly homesick, and behind academically. You see, my plebe class started on March 1, 1910. I've never quite figured out why our plebe year was fifteen months long, but it was. But since we started before the end of my senior year in high school, I was behind a lot of the boys from East Coast prep schools who had taken extra time to prepare. But I caught up pretty soon.

"We competed fiercely with one another, in academics and in athletics. Fiercely, but fairly. We knew the rules: no lying, cheating or stealing. We could incur no debts, not even from cards. No smoking—not a clean habit. Girlfriends, yes, but no one was allowed to marry. Grades every day, every class, every activity. Got so you didn't even notice the scrutiny you were under with the officers watching, their notebooks filled with numbers and letters. Grades, demerits, and class rankings were posted for all to see every quarter. Everyone knew each other's strengths and weaknesses."

"Did people make fun of you?"

Granddaddy responded, "No, but they called me 'Swede,' because my name was really spelled Andersen, the Scandinavian way of spelling Anderson. Administration at West Point changed the spelling to Anderson, probably because it was more English than Andersen, which might have been thought German. But the nickname 'Swede' stuck for a while until people tired of it. Calling me 'Andy' worked out better; that's what my Army pals still call me."

He added, "After the Spanish–American War, the Philippines became part of the United States. One of my classmates was Vincente Lim, the first Filipino to attend West Point. We nicknamed him 'Cannibal.' A proud man who kept pretty much to himself, he was a great Army officer, rising to the rank of brigadier general.

"After Plebe year, all the hazing and razzing stopped. We were now officially part of the Corps. Because we had men from all around the country, we each had different strengths and weaknesses. My strength was horses. Since I'd grown up on a farm, I knew how to care for livestock; many of the boys were East Coast kids, never around a horse or mule. Being an Iowa farm boy, I knew a mule could kick; hit you right, it could break your leg. In those days at West Point, everyone had to learn to ride horses. It was considered part of being an Army officer, and a gentleman. In winter, we rode daily in an indoor arena. And I could ride with the best of them. So, no, even though I was from a small Iowa farm town, my classmates were respectful enough—and so was I. It didn't get you very far to put on airs."

"In my senior year, 1913, I took care of the Army mule at the

Army–Navy Game. The mule is our official mascot. I stayed with him on the sidelines. I didn't ride him, just kept him on a lead. These days, sometimes you see a cadet dressed up as a Black Knight riding around on the sidelines. We didn't do that then. I think there's a picture of me with the mule in my yearbook."

"May I see it?" I was using "may," just like my third grade teacher Ella Thompson said I should.

"Sure, Michael. You'll have to get it off the shelf in the office. On second thought, let me go get it. It's not in the best shape, beat up after so many years."

As Granddaddy leafed through the pages of the 1914 *Howitzer*, he stopped. "Look, here I am on the rifle range. I wasn't a great shot, not like my classmate Vincente Lim, who was an expert shot. I shot Marksman. But my exploits inspired a cartoon. There's the little photo of me with the Army mule in the yearbook. Look here, it's small and out of focus, but that's me." He pointed out a blurry picture; if he said it was him, I'd have to take his word.

"It was a cloudy, cold November day in Philadelphia, where the Army–Navy game is still played. We had lost three years in a row, a real blow to the prestige of the Corps. But that game, we won, 22–9. We were high as a kite, busting-our-buttons proud. We had a great quarterback, Prichard, and an All-American receiver, Merillat."

Granddaddy was on a roll now.

"Merillat was the quarterback; Prichard was his primary receiver. In an age where 'three yards and a cloud of dust' were the norm, the Army passing game took opponents by surprise. And so, in the autumn of 1913, the Army team was undefeated."

He added, "After his Army service, Merillat played for the Canton Bulldogs, alongside Jim Thorpe, in the original National Football League, founded in a car dealership in Canton in 1920."

Granddaddy then asked me, "Michael, do you know the name 'Eisenhower'?"

By that time, I had heard the name a number of times, both

at home and at Nanny and Granddaddy's house. "Yes, sir, I do," I replied.

"To remind you, he was President of the United States before Kennedy. He was also a class behind me at West Point. He was on the sidelines for that 1913 Army–Navy game; he was an Army cheerleader. He'd rather have been playing football, but he hurt his knee badly the year before. He was a good player. Tackled Jim Thorpe one game, when Army played the Carlisle Indian School."

"Did you know any other famous people?" I asked.

"Sure did, and do. My class, and all the classes at West Point, were hand-picked by governors, senators, and congressmen. We were considered potential 'leaders of men.' We were put through pretty hard training designed to make us officers and gentlemen. We were considered elites, ready to do the work of the country, both military and otherwise. Not just West Point, but Annapolis, the Naval Academy, too, like your father."

"I didn't know that. And did you guys stay friends?"

"Yes, my classmates and I were friends, and to this day, we remain friends, the best of friends, friends to last a lifetime. Like my roommate, Charles M. Milliken. For four years, we lived side by side, through hot summers and cold winters at West Point. I heard from Charlie Milliken recently, out of the blue. He wanted to be sure I was coming to the 50th reunion of the West Point Class of 1914. Yes, friends to last a lifetime.

"We did go our separate ways after West Point, seeing one another from time to time, in Washington or the Philippines, or China. We were all working our way up the ladder, seeking good assignments, studying hard, and preparing ourselves in case of war. Good assignments were command of troops, not just schools or staff assignments in Washington. When World War Two came, we were ready, a small group of regular Army officers, well-trained and dedicated to a set of core values. No lying, cheating, or stealing. Those were the rules we lived and worked by, the foundation of the officer core of the largest army this country has ever put together."

The blank look on my young face stopped Granddaddy's story. I was struggling to comprehend the notion of a 50th class reunion.

"Michael," he said, "it'll make more sense later on."

"Granddaddy, what happened after West Point?" I asked.

"I'd say I was a lucky son-of-a-gun. At West Point, I did well, graduating in the top third of my class. I selected field artillery as my service branch. The top 10% usually chose engineering. Field artillery was considered a choice selection. But I still had little direction, until graduation day.

"On that warm June day on the Hudson, I met a great man, General "Black Jack" Pershing. A real Army officer, ramrod straight, handsome with piercing dark eyes. He was the uncle of one of my classmates and was looking for men to join him in a mission along the Texas–Mexico border, the Pancho Villa Expedition. After all the cold and snow of West Point, I jumped at the chance, and before I knew it, my classmate Bill Houghton and I were reporting for duty at the general's headquarters in Texas.

"Not at all like the spit and polish of West Point, the building was an adobe building, one story tall, with hard-packed dirt floors. When we reported, the general welcomed us in warmly, not like the austere man we had met just weeks before. He gave us something to eat, and then off we went to the front, to 'Camp Dust' they called it. Houghton and I rode in the general's car, a Cadillac. It was the first time I had ever ridden in a car, bouncing along with General Pershing. We covered the 15 miles to Camp Dust in a matter of minutes, a distance that would have taken all afternoon on horseback. As we crested the hill to the fort, an 11-gun salute thundered and rolled over the hillside. Apparently, the men were ready for the general's arrival. Of course, I joked that the salute was for the two shave-tail lieutenants newly graduated from West Point!"

"Granddaddy, did you ever receive a cannon salute?"

"Michael, I did, many years later; the one I remember best was the 13-round salute I received in 1943 from the men of the Ozark division, the 102nd Infantry Division I put together from scratch. We started out as a completely green group in the summer of 1942. The men were a mix of draftees and volunteers, from all the states of the union. With a few regular Army officers, I was able to bring them up to proficiency, so good in fact that men were siphoned off as 'levies' to start other divisions. Seemed just about the time I had them sharpened to a razor's edge, off went another group to start another division.

"Anyway, when I got promoted to command of the XVI Corps in December 1943, General Keating took over the 102nd. I had poured my heart and soul into training those men for war. Now I was moving on. After my final speech to the men, they gave me a 13-gun salute. Meant a lot to me; and later, in Europe, they proved up to the task. I'll tell you about them some other time."

"What else happened in Texas, on the border chasing Pancho Villa?" I asked.

Granddaddy replied, "Lots of things, and lotsa nothing. By that, I mean we were bored out of our minds most of the time. It was hot, dusty, and miserable. The horses would get sick from drinking what little water we could find; the water was brackish and alkaline, terrible stuff, but it was all we had for them. And we got sick too—cough, flu, pneumonia, and dysentery. Just a miserable part of the world. The expeditionary force that went into Mexico in 1916 to chase Pancho Villa had a heck of a time; I think they were lucky to get out alive.

"And so was I. One day, my friend Captain John Starkey and I were sitting on a caisson near Douglas, Arizona, talking away about nothing, when a Mexican shell lands about ten feet from us. Knocks us off the caisson; must have exploded, I just can't recall. But neither of us was hurt, just bruised our egos. The men got a kick out of it. Loved that John Starkey, he just laughed it off. But later in France in World War One, he was all business—I'll tell you about that some other time. That night, Grace was glad to see me in one piece."

"Grace, who's she?" I asked. It was getting dark, and I was getting hungry. Granddaddy didn't seem to care. Another bourbon, and he was off to the races.

"So, Michael, do you know about Grace?" he asked.

"Nope," I replied.

Fortunately, Nanny had gone out to make supper. Granddaddy said, "Grace was my first wife, before Nanny."

"What? Granddaddy, you were married before? Did she die?" I blurted out.

"No, she didn't die, she divorced me. Nanny doesn't really like me to talk about it, but it's an important part of my life. So keep it a secret, OK? Our little secret."

I nodded, wanting to know more, not aware that keeping secrets is difficult, and at times destructive. "So what happened?" I asked.

"When I first arrived in El Paso, General Pershing was invited to every club, town hall, and gathering in town. After a while, it was just too much for him. He had lost his wife and three daughters in a house fire in California six months before and was in no mood to go out and about, entertaining and being entertained. I think he just wanted to stay home with his son, his only remaining child. So, instead, Bill Houghton and I went, to represent Pershing and the U.S. Army.

"We were 'officers and gentlemen,' trained in dance and etiquette at West Point; really, we were. We had dance classes every week. And Bill and I were good at it. We attended balls and dances at the El Paso Country Club almost every month. And on one of those days, I met Grace, Grace Amoleyetto Wingo, the 18-year-old daughter of an El Paso banker, and railroad man. She was great fun, vivacious, and a great dancer. She had me hook, line, and sinker—and she knew it. Before long, I was turning down all the invitations, unless Grace was included. And six months later, we were married, spring 1915. I was 24, she was still 18. What fun we had, swimming in the Rio

Grande, carrying on. I took her north to Iowa the following Christmas to meet my father and sisters. They all thought she was grand, too, bundled up against the winter winds, flashing her wedding ring for all to see.

"But the marriage couldn't stand up to what else was going on in the world, and my small part in the larger drama," he added. "While we were chasing Pancho Villa hither and yon, the war in Europe raged on. The Germans were encouraging the Mexicans to stir up trouble along the border. Germany did not want us to help the English or the French. They really didn't want us to get in the war at all. The Germans declared that all ships coming into England or France would be sunk, whether they were military or civilian ships. In 1915 the Germans torpedoed a British passenger ship, SS *Lusitania*. Among the hundreds of victims were American citizens. The sinking stirred up bad feelings against the Germans in America, but they apologized and promised no more torpedo attacks on civilian ships. And for nearly two years, they kept their promise, even though civilian ships were carrying guns, ammunition, and essential supplies from America to the Allies. Instead of sinking those ships, the Germans stirred up trouble in Mexico, thinking America would keep war material at home, instead of selling it to England or France.

"The Germans shipped modern guns and artillery pieces to Pancho Villa and encouraged him to attack across the border. For a while, the policy worked. We were preoccupied by his attacks, and we stayed out of the war in Europe. Revolution was still sputtering in Mexico, and the State Department secured evidence indicating that Germans were aiding and encouraging the activities of the factions in Mexico, the purpose being to keep the United States occupied along its borders and to prevent the exportation of munitions of war to the Allies. But the Germans pushed it too far.

"In 1917, the German foreign minister sent a confidential note, the infamous Zimmermann Telegram, promising full recognition of Mexican claims to any territories Mexico might take from the

United States. In other words, Germany wanted a full-blown war on our southern border, another Mexican War, with battles like the Alamo. At the same time, Germany restarted unrestricted submarine warfare, sinking any ships that entered British waters. All of this was too much, too provocative for President Wilson, passionately opposed to military solutions to political problems. Even Wilson recognized we had been forced into a corner."

Granddaddy continued his part in the story, "Before we knew it, President Wilson declared war—for good reasons, but the fact is, we were unprepared. And Grace knew it, too. She had read the newspapers where hundreds of men were killed every day on the Western Front. She pleaded with me to get out of the Army, to stay in El Paso where her family was, where it was warm. She was so young and had no experience with the U.S. Army. She couldn't understand how I had to put duty and country before her. She couldn't understand that I was honor-bound as a West Point man to obey the Army, to go wherever they sent me. She was so angry, and scared. She knew I might not come back from France.

"But my duty was clear. I was committed to the Army, to my brothers from West Point. I'd never turn my back on them. The U.S. Army was about to take on the largest, most professional fighting force in the world, the German army, which was beating the British and French in the west, the Russians in the east, and the Italians in the south. I had to go.

"I left for France in July 1917, part of the first wave of American troops under General Pershing's command, the Big Red One. In my pocket was a letter from Grace, a letter she had made me promise not to read until at sea, on my way to France. We pulled away from the pier in Hoboken, New Jersey, and headed out to sea, gliding past the Statue of Liberty.

"I was overwhelmed by the vast ocean, a sight I had never experienced before. Of course, I had stood on the banks of the Hudson River at West Point before, I had even stood above wheatfields and cornfields in Iowa and imagined the ocean must look like this. But nothing compares to the vast expense of sea, stretching over the horizon, clouds touching the waves, waves banging into the ship, tossing us about. Would I ever get home, I wondered? Would I see the Statue of Liberty again, or the fields of West Point just up the Hudson?"

"Later, I read Grace's 'Dear John' letter. She wanted a divorce. She was not going to wait for me to come home."

Nanny called from the kitchen, "Dinner's ready. Come and eat before it gets cold."

Granddaddy just sat there, looking off in the distance. I'm not sure he heard Nanny's call. He sometimes "played deaf," blaming his time with the field artillery. Maybe he was deep in thought about the olden days, or maybe he was just ignoring her. Regardless, I was hungry.

"Granddaddy, I hear Nanny calling. It's time for dinner. I'm sorry about Grace. I'll tell no one about her. But I think we'd better go eat dinner."

And until now, I've never told anyone about Grace. Granddaddy and I had a secret, something only he and I shared. Or so I thought. You see, the thing about secrets is the fact that most everyone is already in on "the secret." Grace was no secret, but a person about whom neither Nanny, my mother, nor Granddaddy would speak openly. A long ago marriage that ended in divorce was still considered a shameful and humiliating experience, a failing not fit to discuss in polite company.

How sad it is to me now that Granddaddy's experience could not be discussed during my lifetime. So what if his first wife sent him a Dear John letter on his way to France. He didn't do anything wrong. Maybe it really was for the best. Maybe Grace and he were not compatible? Who knows? I certainly don't know, because it was a secret.

I regret never having the opportunity to discuss this or other adult problems with either my grandparents or parents. I was always the six-year-old who needed to be protected. If I learned that everyone made mistakes and that perfection is rare, I think they thought maybe I wouldn't try very hard. The message to me was that I was expected to be well-mannered, to get straight A's, and to be undefeated on the athletic field.

When I came in second in a first-grade election, I cried bitter tears. Tears of shame for losing, and tears of fear for what might happen to me. When third-grade addition answers were wrong, I covered up my mistakes out of fear of being mocked. And later in life, when my own marriage ended in divorce, I didn't know how to reconcile my vows with my failings.

My parents were each only children, so no aunts or uncles were around to tell stories of their lives. In addition, I had only one set of grandparents, Nanny and Granddaddy. I never knew my father's parents; my paternal grandmother Ethel Stover Van Ness died in 1919, four months after my father's birth. Her husband, Harper E. Van Ness, my father's father, left him to be raised by his mother, Annie Harper Rixey Van Ness, my father's grandmother. Annie had two other children, Pauline and Gladys, who acted as surrogate mothers to my father, and whom I came to know as Aunt Pauline and Aunt Gladys. They both called my father "son." When Harper E. Van Ness remarried, my father did not go to live with his father, but stayed with Annie, Pauline, and Gladys. When Pauline married Herschel Schooley, Mr. Schooley became a surrogate father to my dad. My brothers and I called Mr. Schooley "Uncle Hersch." This was all very confusing to me as a youngster, especially since the relationships were never fully explained.

How much healthier our relationships might have been! How much better to understand, discuss, and learn from the foibles of one's ancestors. Instead of a bitter silence or a wall of shame, we could have learned so much in the acknowledgement of mistakes of the past. Instead, we focused our family efforts on striving and achievement. Goal setting can be valuable. Achieving goals can be fulfilling. Unfortunately, all too often in the wake of significant accomplishment lay quiet despair.

For six months, Granddaddy despaired at the loss of Grace, the headstrong 18-year-old Texas woman who had stolen his heart. Then he began a diary of his experiences in the trenches of France. From his writing, you would never know his heart had been broken.

4

A Pile of Manure

Sitting in the living room and preparing to smoke, Granddaddy continued his storytelling a week later. "When we arrived in France, we were treated like conquering heroes. As we pulled alongside the pier, our band broke into 'The Marseillaise' and the cheering crowd roared themselves hoarse. But as we disembarked, it didn't take long to see that the French were in trouble. Everyone was dressed in black, mourning the loss of a son, husband, brother, or more. There were many invalid Frenchmen lining the streets, on crutches or with empty sleeves. They watched us with pity in their eyes, knowing full well our fate at the front. Nearby, we spied a group of German prisoners, hearty fellows in heavy wool uniforms. They appeared not too sad; in fact, as we stared, some smiled, likely happy to be out of the fight.

"I couldn't show my true feelings. Officers had to keep up morale, keep a 'stiff upper lip,' but I was worried inside. What had we gotten ourselves into? The United States lacked all the tools of war in 1917, except manpower, and the Entente needed manpower. So, instead of bringing all the equipment needed for operations in the field, like horses, cannon, field kitchens, and wagons, the Sixth Field Artillery left all that behind in America, receiving French equipment upon arrival in Europe.

"The French army did their best for us, feeding and sheltering us along the way. But it wasn't what we were used to. For instance, we were herded onto the French trains for transport from St. Nazaire to Besancon, in eastern France. Each train had 30 to 40 boxcars for the enlisted men and two passenger cars for the officers. While the officers had the benefit of passenger cars with benches and fold-down beds, the 40 enlisted men were crammed into windowless, seatless boxcars, designed to carry eight horses. These were stables, not passenger cars—hot, stuffy, and poorly ventilated—the type of prison cars that would be used to transport Jews to concentration camps in the next war. We made sure to stop frequently, to get the men out for coffee and rations, to stretch, and use the latrines. Even so, it was a miserable trip for the men.

"General Pershing had arranged for us to be sent to a quiet sector of the Western Front. We needed to recover from our journey from the States, to exercise the men and get fit enough for combat, and to train on the French 75mm cannon. As I recorded in my diary, we started work with a will."

"A diary," I asked. "You kept a diary?"

"Of course," Granddaddy replied. "Everyone kept a diary. It was a way to pass the time, to record your impressions of people and places."

When he showed it to me, the diary looked like a school workbook, not a leather-covered book with "My Diary" embossed in gold. On the cardboard cover in Granddaddy's cursive were the words, "Diary of John B. Anderson, Captain 6th Field Artillery, U.S. Army. In case of my death, please send this diary to Mrs. Henry Johnson, Jr., P.O. Box #27, Parkersburg, Iowa."

Granddaddy sat back, knocked the ash out of his pipe, and filled the bowl anew with Prince Albert tobacco. His face clouded over, his eyes grew misty, his throat hoarse. Then he began to read:

> My first experience at the front. Only one battalion from
> each artillery and infantry regiment was to be at the front at

a time. The French had control of us. We were to do nothing, no firing, no fighting without their consent. Since there was sort of a "gentleman's agreement" between the French and the Germans to keep this part of the western front quiet, the French didn't want eager Americans to spoil things, to stir up unnecessary trouble.

But the French couldn't hold us back forever. In fact, within weeks, our first battalion, straining at the bit to get at the Germans, left for Nancy, situated along the Rhine-Marne canal. The battalion consisted of Batteries A, B, and C, and was commanded by my old friend, now Major John Starkey—you remember, my friend who was blown off the caisson with me in Arizona.

With the consent of the French, our men labored all night, got our gun in place, and then realized – we had no ammo! The French controlled everything, including the distribution of shells, and had refused to give any to the Americans, that is until Captain McLendon of Battery C convinced the French Commandant, Major Villers, to give us some.

How he did that, he told us later, was to appeal to the Major's sense of patriotism, American and Frenchmen fighting shoulder to shoulder against a common tyrant, new world crusaders of 1917 who heard the same call that brought Lafayette to our shores in our Revolution in 1776. And besides, this American battery was a part of his own battalion, fighting on the sacred soil of France. He'd share the glory too. We got our ammunition.

Sergeant Arch was the Chief of Section (Acting No. 1) of Battery 'C' for the first shot fired into German lines. On exactly 6:05 a.m. on October 23, 1917. The lanyard was pulled by Corporal Osborne de Varila. America was in the war, and the Sixth Field Artillery fired the first shot!

The shell casing was sent to President Wilson as a souvenir. Who knows where it is now. I hope someone didn't just throw it away. Anyway, the Sixth stayed in the line for the next two weeks, firing from time to time, re-positioning the guns to avoid counter-battery fire from the Germans.

After two weeks in the line, Battery C returned to winter quarters. As the unit moved away from the line, they found their path marked by fresh flowers in almost every village, a tribute from the French for their efforts. Having fired the first shot, they were famous. Beating the infantry to the punch was a great source of pride to the field artillery. Major Starkey was to say, "the infantry beat us to France by nearly two months, but we of the artillery got into action more than twelve hours ahead of the Doughboys" [a nickname for American infantrymen]. The French 75mm cannon used for the first shot was carried home and now sits at West Point, a fitting reminder of the role they played in World War One. Cadets paint it from time to time, in outrageous colors meant to inspire the football team, or to celebrate a victory over the Navy.

About this time, Nanny came in. She asked, "Did you tell Michael about the manure pile?"

Granddaddy smiled. "Ah, yes, the manure pile. Saved my life." He began to read aloud again:

My Colonel, W.S. McNair, was promoted to brigadier general and took me for his aide de-camp. We took many trips in his automobile, and I saw a lot of that particular section of France. Besancon, for instance, is a very interesting old place. It was captured by Caesar in his time, and many of the old Roman ruins still remain. We went to see them, and they were well worth the trip. Also, we went to the Swiss border, and though we could not cross, yet I am sure I had at least one foot on Swiss territory. We went to the source of the Loire, which is a river which flows out of the side of a hill, through a very deep gorge – almost as wonderful as our own Grand Canyon. Also, we went to see the "Glacier," which is a deep cave – about 200 feet deep – in which ice forms the year round.

Not much of a manure story, I thought, impatient for something better. His experiences seem more like those of a tourist, not an American soldier in France to rid the country of the dreaded "Boche."

Granddaddy continued to read:

> Joe Swing, Class of 1915, and in the Field Artillery, was aide to General March and we use to go off on motorcycle trips also to various places. It was on one of these trips that we hit an ox-cart and were both thrown onto a manure pile. We were knocked unconscious, out cold, and were brought home by a peasant. I came to 12 hours later and had no more idea than a rabbit as to what had happened. The motorcycle was unhurt. You can imagine the odor which persisted in staying with me for weeks.

Nanny chimed in, "That pile of manure probably saved your life!"

"I suppose so," Granddaddy replied softly. He wanted to keep reading. "The next page explains more about French villages, and farms, and manure piles," he said.

> The country is intensely cultivated, except the wooded areas, of course, which are very numerous and extensive. There are very few isolated farms, as all the farmers gather in small villages, which cover the country. The villages at a distance—with their massive churches with tall spires, always present—are very picturesque, but at closer quarters they are filthy, smelly places, with cows, chickens, pigs and sheep filling the streets. Also, the manure heaps fill half the street, for among the peasant class, a man's wealth is judged from the size of the manure pile which adorns his portion of the street. (But I'll not complain about them, for, as you saw in the preceding page one of them doubtless saved my worthless neck from being broken.) Also, the stable is always in the same building in which the human beings of the family live.

Granddaddy stopped, filled his pipe, and lit up. "So there it is, the manure story. Glad I survived," he chuckled.

Although I was to hear the manure story from various family members throughout my life, hearing it from Granddaddy, read directly from his diary, was the best. It was also one of the few times that he smiled. We all chuckled, for as we all know, boys love scatological stories.

Even today, I love the story, but for a different reason. I can visualize Granddaddy and his friend careening around the corner, without a care in the world. I can identify with him. I was once that young officer. My exploits were confined to hanging out of a helicopter on a wire rope rather than skidding around a corner. Both he and I were carefree, invincible, and looking for adventure. I like the fact that we both had friends we admired who were destined for greatness. Granddaddy's friend Joe Swing became a fabled airborne general leading his troops in battle in the Pacific War. He ultimately made lieutenant general, served as Commandant of the Army War College, and commanded the U.S. Sixth Army.

For my part, I was fortunate to meet my fair share of heroes during my own military service. At Bethesda Naval Hospital, one of my intern colleagues was naval aviator LCDR William Shankel. He was older than the rest of us, very quiet and very serious. At times he seemed preoccupied with matters beyond the usual chores of medical trainees. No wonder: he had endured six years of captivity in North Vietnam during the Vietnam War before going to medical school and pursuing a career in surgery. I could only imagine the hardships he endured.

More outgoing was another intern, Dr. Carl June. He was a 1975 graduate of the U.S. Naval Academy, number two in rank in the class my father had hoped I would enter. We played tennis at times; but more importantly, he was a brilliant physician who went on to become a leading expert on gene-splicing and is now Director of the Center of Cellular Immunotherapies at the University of Pennsylvania. If he is awarded a Nobel Prize in Medicine, I won't be surprised.

It was past dinnertime; Nanny reminded us that she had dinner on the table. Meatballs doused in Lawry's Season Salt and baked potatoes with whipped butter waited for us in the dining room. As always, there was good china and a heavy silver knife, fork, and spoon at each place setting. We didn't say grace; when Granddaddy started to eat, so did I.

During dinner, I asked, "Granddaddy, sounds like you had a lot of fun in the Army, driving around, seeing the sights in France."

Granddaddy was chewing, so Nanny took the lead. "Michael," she said, "for me, life in the Army was both great and terrible. Great friendships, travel around the world, and always something new. But the pay was terrible, and promotions were awfully slow. Sometimes someone got promoted ahead of you, and you just didn't know why."

"How about during the war when Granddaddy was overseas? What did you do?"

She replied, "Well, Michael, we weren't married until after World War One. But during World War Two, I suffered so. I worried myself sick that something bad would happen to him. It was very lonely, too. I wrote letters to your grandfather almost every day, and sometimes, it'd be a week before I received any mail in return. Other days, three or four letters would arrive. I just fretted so that he would get sick, or wounded, or worse. I knew of officers, West Point men, who were in plane crashes, or lost at sea, or captured by the Germans. It was torture for me, at times, really."

Granddaddy piped up, "Sue, you were a trooper, not just when I was overseas, but whenever we had to change duty stations, sometimes with very little notice, like after the attack on Pearl Harbor in 1941. I could always count on you to take care of little Tudy and the house."

Nanny had tears in her eyes. Seeing her cry made me very uncomfortable. Crying was discouraged at home. My mother, also named Sue but who they called "little Tudy," never liked displays of emotion, especially tears. I remember the time I announced proudly that I had not cried once that day; I was about five years old. And that's the way it was in our family. If something was uncomfortable or painful, it was better to avoid talking about it.

Later, as I drifted off to sleep, I tried to forget about Nanny's sadness. Instead I imagined the excitement of life at the front in World War One, riding a motorcycle around the French country-side, visiting Roman ruins, and shooting guns at the Germans. Like most young men, I put the descriptions of rain, and mud, and cold out of my mind. I ignored the fact that every ounce of the enemy's energy would be used to devise ways to kill, or maim, or incapacitate me and my friends. I ignored the fact that doctors could only fix so much. In my young mind, being wounded was more likely an oppor-tunity to flirt with a pretty nurse in a bleached, starched uniform, not a time of pain and suffering.

I thought bad stuff happened to the other guy. I was invincible, like Granddaddy. Hadn't he gone to war, had fun, and come home without a scratch? I had no idea that some wounds were invisible, like the festering memories of lost friends or the shameful acts of cowardice, theft, and greed that men commit at the end of their rope. I could not conceive of anything bad happening to me. Instead, I saw the uniforms, helmets, and medals. I dreamt only of the glory of it all as I thought, "Maybe I'll join the Army someday. Fire the first shot, crash a motorcycle into a pile of manure. I could be the first man over the top, leading my men to victory, and come home without a scratch, to Nanny's meatballs and baked potatoes."

Later, I'd learn that veterans of war suffer wounds both visi-ble and invisible. During my time at Bethesda Naval Hospital, I witnessed a Marine Corps major, an aviator whose right arm was severed at the elbow by a rocket grenade, weep in despair knowing he would never return to his unit in Grenada. I also cared for men and women whose memories of combat in the Pacific, of piles of human teeth from victims at the concentration camp Ebensee, and of friends who never returned plunged them deep into their drinks, with alcohol abuse ruining their health and family relations.

5

Rock Cookies

The next time I slept over at Nanny and Granddaddy's house, I wanted to know more about his World War One experiences.

"Granddaddy, what was it like in France?"

"Michael, it's hard to say. For me, it was wet, cold, and muddy."

"But were you ever really scared?"

He responded, "Yes, I was. But at those times, like being shelled or gassed, I was so busy I hardly had time to think. Later, when things would calm down, that's when I'd get scared, thinking of all the things that might have happened." He paused, and then added, "As funny as it sounds, when it was scary, you wanted the scary stuff to stop. But then, you wanted it to start again. Kinda like a roller coaster, or a really scary movie."

I was puzzled. "You mean, like a horror movie?" While I liked roller coasters, Dracula movies were too scary.

Granddaddy replied, "Yeah, kinda like a horror movie." He paused. There was no way to make a seven-year-old boy understand.

"Wanna see my helmet from the war?"

Of course, I did; and anything else he wanted to show me. So out came the helmet, and gas masks, and leggings. Wool shirts and scruffy wool knit caps.

"When I first arrived in France, other than the first shot into the

German lines, I spent most of my time in school, at Langres, a sort of finishing school for young officers to learn staff work. Just classroom work, not really what I thought war was all about. And mainly in French, with French instructors, for French equipment.

"One day, after another boring lecture about trench warfare, an English officer came by. I don't remember the subject but at least I could understand most of what he was saying. And then he inquired about volunteers. 'Would anyone like to spend a few days with us, up in Flanders?'

"I leapt at the chance. Where we were in France, near the Swiss border, and it was cold, really cold, and boring. I didn't want to sit around in classrooms all day. I wanted to see more of the real war, the real action. Other than the firing of the first shot in October, I had spent nearly seven months in France studying, and training, and learning to fight—but not doing any fighting. Next thing I knew, I was on a train going to Ypres, site of the British front lines, site of several bloody battles over the previous three years, but in February 1918, a relatively quiet zone.

"I was attached to the British XXII Corps, commanded by a General Godley of New Zealand, as an observer. I was flattered to be asked to dinner with the general. He was a bit standoffish, like many British aristocracy. He told me that he'd visited the States; in fact, inspected the West Point Corps of Cadets in the autumn of 1910. Unfortunately, I had to admit no recollection of the event. I don't think he liked that much. After my reply, he directed his attention to the other men at the table. That was fine by me.

"To tell the truth, the British were courting us, trying every trick in the book to get us to commit troops to their depleted divisions and regiments, under their flag. When I asked to go up to the front, they were most accommodating, and I spent time in a listening post in no-man's land, the ground between our trenches and the enemy's. I was also given airplane rides over our lines, to see our positions relative to the Germans'. We never crossed over the line, never flew into areas that might provoke a response. I think the flyers were told to keep me alive. It would not have helped to get me killed.

"Flying was a great thrill; to be up above the mud and stench made me think that flying, instead of field artillery, might be a better way to go. But after my time with the aviators, I returned to my own crowd. What a fine bunch of men, true and honest. I put out of my mind any thoughts of changing to the air; even if I found myself under counter-battery fire or in shell holes and trenches, the field artillery was home for me.

"Michael, see the white dirt on my helmet? That's the chalky soil of France, from the trenches. After a night in a British heavy artillery headquarters, a house of sandbags, it was back to the gun batteries. There, I visited their flash-spotting and sound-ranging stations, men trying to locate the enemy guns, so we could return fire at them. The work was nerve-wracking. Get the location of the Germans right, and the shooting stopped. Miscalculate, and the Germans then knew exactly where you were, and poured it on with a vengeance.

"I only stayed in Ypres for two weeks. I was needed back with our own forces. Our numbers were growing by leaps and bounds, thousands of Doughboys arriving weekly. Officers like me were needed to train and organize these newly arrived men. It was a race between the reinforcements from America and the German soldiers arriving from Russia.

"The Germans were planning a major offensive in the west for the spring of 1918. They calculated that the Americans were too inexperienced to provide any real help. The English and French were nearly 'bled white' from three years of war. Millions had died and no replacements were available, not in the numbers needed to withstand the Germans. With the collapse of the Czarist Army, the Germans brought hundreds of thousands of experienced troops to Flanders and France, for a final big push. It was up to us, the United States, to save the day. It was time for me to say goodbye to my new British friends and get back to my men. The fates of these British officers, I never found out. A few weeks later, the Germans hit them, pushing them back toward the French border. All I can say—brave lads, all."

When Granddaddy stopped talking, I put on his helmet. It didn't fit, was way too large. Even the chinstrap couldn't keep it from wobbling around. Taking it off, I then tried his gasmask. In fact, there were two of them. Granddaddy said one was French, for mustard gas; the other was British, for phosgene. I thought they were nothing but stuffy and smelly. I couldn't see out of the cloudy plastic lens. I couldn't breathe either. The hood was dirty, stiff, and uncomfortable. Even at that age, I could tell that gasmasks were awful.

"Michael, you're too little for those things. Put them back in their pouches."

I crammed them back in, folded over the flap, and tightened down the retaining strap. I slumped into my chair, glad to be done with Granddaddy's helmet and gas masks. They must have been a lot better new, better than forty or so years old.

"Tell me about your day at school," he asked.

I brightened, "We had riflery today. Since it was too cold and snowy for athletics today, we went to the rifle range. Shot .22 rifles, from the prone, sitting, kneeling, and standing positions. For the prone position, we lay on old camp mattresses, thick cotton ones with blue stripes. Then we rolled them up and shot the other stances. Targets were 50 feet away, not 15 feet like at camp. Had to squint to see through the sights, circles on the end of the barrel."

Granddaddy asked, "How'd you do?"

"Ok," I replied. "I did what you told me to do. Line up the bullseye, breathe out, and slowly squeeze the trigger. Afterwards, the instructor asked me, 'Where'd you learn to shoot?'

"I answered, 'In California, from the Marines.'

"He added, 'Well, they taught you well. You shot 90 out of a 100. Very good; in fact, expert.'

"I told him, 'I also got some pointers from my grandfather. He went to West Point.'"

Granddaddy sat quietly a moment, taking in the pleasure of hearing about my exploit and his part in my success. Knowing my attention span was short, he asked me to get the chips; we had time for a few hands before dinner. As I scurried about, getting the cards and chips, putting the helmet and gas masks away, Granddaddy poured himself a little snoot of bourbon to ease himself into the evening's shadows.

"Granddaddy, how'd you get home from the war?" I asked.

"Shortly after my time with the British, I commanded a battalion of the 6th Field Artillery. We were hit hard by the Germans, the battalion commander, a major, was overcome with gas and one of his batteries wiped out. Since I was an adjutant to our commanding officer, General Summerall, I was put in charge of the 2nd Battalion, even though I was only a captain. Got things right, repositioned the guns and got them working again. A month or so later, the major had recovered and returned to the battalion, so I was out of a job."

"At the time, April of 1918, it was thought that the Americans would need another year to get ready to be set to take on the German Army. Our forces were growing in leaps and bounds, but there weren't enough trained officers to command the men. Pershing had tried to set up schools in France to do the job, but it wasn't working out. So it was decided that training of officers should occur in the United States before they were sent to France. And it was thought the training would best be done by guys like me, with experience in the field, time in the trenches, men who'd been under gas attack and bombardment. Since I was an artillery officer, it was decided that I would go to the Advanced Artillery School at Fort Sill for further training. Then I'd be ready to teach all the officers before they came over.

"I didn't want to leave. I didn't want to leave my men, my unit.

I felt like I was letting them down. I felt a part of something bigger than myself. I was meant to be in the fight. My objections were noted. In fact, General Summerall, Commanding General of the First Artillery Brigade of the First Division, agreed with me that I should stay. The Brigade and the Division were preparing for an attack on the Germans at Cantigny, the first offensive action by an independent American force. Seemed to us that it would be best to have experienced officers involved.

"But General Bullard, the Commanding General of the First Division, was under orders to send a quota of officers, a set number, back for this duty. Since the major I had temporarily filled in for was recovered and back with his unit, I was available. It would have been fruitless to object further; in fact, it would likely have caused trouble. General Pershing was not one to cross. Better to grin and bear it, do one's duty.

"On April 24th, 1918, Pershing gathered all of us together to give a pep talk before the planned attack at Cantigny, the first offensive action by American forces in the war. My heart was heavy. I knew I would not be part of the attack. I felt like a bit of a fraud. The upcoming trial by fire would fall on my fellow officers, not me.

"I remember the day well. Pershing's words are still clear in my mind:

> We Doughboys represent a young and aggressive nation, the mightiest nation engaged in the war, there to defend the sacred principles of human liberty on European soil. Now is the time to repay our Allies for their patience, and to make our enemy understand the true power of the American fighting man."

Pershing's speech was called "Pershing's Farewell to the First," for the First Division was now under the command of the French, although its internal command structure was American. It was Pershing's great accomplishment that American troops were not sent as replacements piecemeal to French and British units.

Granddaddy continued:

"I'll say this. When I got home to Iowa, my sisters and my nephew were sure glad to see me. Since I had been without leave for over a year, I was given two weeks' time to visit Parkersburg. My sister Margaret, 'Mrs. Robert Johnson' on the diary cover, was particularly happy. She made my favorite cookies, the ones we called "rock cookies" because they got so hard after a few days. I showed them this same helmet you're holding. Gave a little speech in my hometown. Got my name in the newspaper. Felt like a hero for a moment or two until I realized that my friends, my men and officers, were in the fight of their lives at Cantigny. Later I learned that over 300 Americans died in that fight.

"Later, in 1921, I was stationed at the War Department, in the Field Artillery branch. While I was there, I was invited to participate in the burial of the Unknown Soldier from the First World War. President Harding invited all branches of the Army to participate, to escort the flag-draped coffin from the U.S. Capitol to Arlington. It was a magnificent tribute to honor all the unknown soldiers who died in France. In comparison to the Battle of Cantigny, more than 20,000 men died in the Meuse–Argonne. Maybe it was just as well that I came home when I did."

I like to remember that day, the discussion of shooting and life in the trenches. It was one of the few times that Granddaddy dropped his gruff exterior and spoke about his own feelings. Whereas he was gung-ho and full of piss and vinegar to get at the Boche, at least in his diary entries, he relished his sister Margaret's "rock cookies" as a reminder of home. More than 100 years later, I can only imagine how much it must have meant to receive a tin of homemade cookies. How many weeks must it have been from the time they were made in Iowa until their arrival in France. No FedEx or airmail in those days. Truly, those cookies must have been as hard as rocks.

I got the recipe for "rock cookies" from my mother. Rich in molasses and ginger, I find them perfect with my afternoon coffee, especially in winter. I also give them out as Christmas gifts. I warn folks, if left out too long, they get hard as rocks.

Here's the recipe:

Aunt Margaret's Rock Cookies

½ cup of white sugar

1 cup butter

½ cup molasses

3 eggs

1 tsp. soda

1 cup raisins

¾ cup nuts (broken pecans or walnuts; coarse, not too fine)

1 tsp. ginger

1 tsp. ground clove

1 tsp. cinnamon

pinch of nutmeg (hearty)

4½ cups of white flour, more as needed

Mix all ingredients together until stiff. If batter flattens at room temperature, add more flour. Drop on cookie sheet by teaspoonful. Bake 12 minutes at 350 degrees. Do NOT overbake.

6

The Shrimp Bowl

With my background as the youngest of three boys in a family steeped in Southern military lore and the school's heritage, my attendance at the Landon School seemed pre-ordained, a natural fit. Nevertheless, I was underprepared and struggled for a while to catch up when I entered the third grade at Landon in the autumn of 1961. As mentioned, I had completed second grade at the local public school, Radnor Elementary. I learned on the fly to add multiple columns of numbers: 63,754,756 + 37,746,386 + 74,739,398 =? The answer is 176,240,540. Most of my classmates came from the local private primary schools, where cursive was taught in second grade. I had to teach myself by copying the letters displayed above the blackboards, always with a fountain pen. At times, the academics were overwhelming; fortunately, there were also athletic opportunities, starting with football in the fall.

The third-grade football team, of which I was a player, practiced all autumn to prepare for its one game, the Shrimp Bowl, played against the 65-pound fourth graders. It was unusual for boys so young to play tackle football, but Mr. Banfield thought it best that boys have "an arduous beginning." The Shrimp Bowl was rarely a fair fight, since the difference in slightly higher average weight of the third graders was insufficient to negate the age advantage and maturity of the lighter, but older fourth graders. Allowing the third graders, regardless of weight, to play the fourth graders in tackle football might seem a bit of a mismatch. But in those days, some of the third graders had been held back before entering the third grade, so they in fact were as big and strong as some of the fourth graders who had been promoted in the usual manner. It was an imperfect system but accepted in the spirit of English schoolboy traditions.

Even with the occasional mismatches in weight across the line of scrimmage, the third-grade team was considered "David" in the "David versus Goliath" contest. We boys didn't know it at the time, but later came to realize that in football, and romance, there is great virtue in unequal odds. Even the presence of third-grade boys who had stayed back or repeated third grade was rarely enough. But the sight of third and fourth graders running around, tackling and blocking one another, was unusual and amusing for all. One year, the game aroused enough excitement that the *Saturday Evening Post* devoted four full-color pages to the contest.

My third-grade year, the game was eagerly anticipated. The year before, in something of an upset, the third-grade team won. Now there was the chance that the third graders would make history. Never in the history of the Shrimp Bowl had the third grade won two years in a row. And so, it was with great hope that we took the gridiron. There was another bit of history being made: Tom Perry, of the fourth-grade team, was playing in his third Shrimp Bowl, the consequence of repeating third grade the year before. As it turned out, Tom Perry would be the star of the game.

Granddaddy came to the game with a couple of West Point buddies, Generals Anthony McAuliffe and "Tooey" Spaatz. All three men met up from time to time at The Army and Navy Club at Farragut Square for drinks and dinner. (It was there that the West Point Club of Washington met to choose its annual award for Landon's outstanding Fifth Form boy.) For them, the outing to the Shrimp Bowl was a chance to get some fresh air, to reminisce, and to see and enjoy the fruits of their wartime labors, young boys growing up free and strong, safe from the Germans and Japanese. While General McAuliffe had no sons at Landon, he had heard about the school from friends and neighbors. General "Tooey" Spaatz' grandson, "Tooey" Thomas, was playing for the third-grade team. Granddaddy was proud of the fact that I was the quarterback, and the team captain. I suspect he invited his two more distinguished and famous friends to show off a bit.

Granddaddy led Generals McAuliffe and Spaatz down the grassy slope toward the varsity football field, which was lush with thick grass, carefully tended and reserved exclusively in those days for varsity football games and the Shrimp Bowl. All other games were played on practice fields. It was a thrill for the third and fourth graders to be playing on the beautifully maintained grass, lined with chalk, and distinguished with two bone-white goalposts at each end.

Granddaddy spied my brother Scott, a First Former (seventh grader) coming to watch the game. Scott was a real history buff, an expert in military affairs. Granddaddy introduced General McAuliffe. Scott recognized his name immediately, the hero of the Battle of the Bulge, the general who responded to the German demand to surrender the encircled town of Bastogne with the famous, one-word reply, "Nuts!" After shaking hands with both McAuliffe and Spaatz, they stood together awhile watching the action on the field.

"Andy, what number is your grandson?" General McAuliffe asked my grandfather, using the nickname, derived from the surname Anderson, that he had acquired during his Army days.

"I'm not quite sure. Hard to tell from this distance. He's the quarterback on offense," he answered.

Spaatz added, "Hard to tell who is who. A bit like a beehive, I'd say."

It was an apt analogy. We were a swarm of young boys, striving to do our best, while providing a sweet moment for the three old warriors watching the next generation follow in their footsteps.

After a bit, French teacher Colonel Hopkins approached. Of course, he knew all three generals, having introduced himself years before during West Point Club of Washington meetings, an old boys' club of military men in a city replete with old boys' clubs. Hopkins was but one of many who exalted in General McAuliffe's exploits and basked in the reflected glory of the defender of Bastogne. General Spaatz was likewise heralded, the leader of the Army Air Corps in World War Two, and in 1947 the first Chief of State of the newly independent U.S. Air Force. Here at Landon, overlooking the young boys playing, the four men had time to talk and share more stories of their lives during and after the war. The older generals expressed particular admiration for Colonel Hopkins' continued career as a teacher at Landon. Of course, Colonel Hopkins reminded them that he was more than ten years younger, and his colonel's pension was smaller than the one given to general officers.

The game was close, but we lost—a crushing blow to my young ego. Since I was team captain and quarterback, I felt a particular responsibility for the loss as I mulled over the details of the game. My good friend Tommy Wadden played halfback. Repeating third grade coming into Landon, he was older, stronger, and faster than most of us. He scored our first touchdown, but we couldn't get the two-point conversion. Then the fourth graders scored on a pass, and Tom Perry ran the ball around right end for the two extra points. Going into the fourth quarter, the third grade trailed 8–6. Then late in the game, Tommy had a long run, a T-7 reverse, down to the two-yard line. First and goal to go. All we needed were two or three yards. We could do it, I thought. Piece of cake. Victory was soon to be ours! I wanted the glory of scoring the winning touchdown. Tommy had already scored so it was my turn, I reasoned. Believe it or not, I called four quarterback sneaks in a row. No plays were sent in by our coach, Mr. Bates. Play calling from the sidelines was rare in those days. Substitutions were also unusual; it was the days of "Iron Men" playing the entire game, offense and defense without a break, a test of determination and manhood. I ignored the pleas of the parents on the sidelines and of the other boys in the huddle. Someone said,

"Another quarterback sneak isn't going to work!"

I ignored him. Nope. I wanted the glory of scoring the winning touchdown. Never got close; in fact, with each play, we lost ground. The final score was 8–6. A few days later, the *Landon News* headline "Third Grade Loses Bravely" rubbed salt in the wound.

When the game ended, my grandfather and the other men who had assembled to watch pledged to stay in touch and see one another before too long, and then they departed for home. It was my turn to spend the night with Nanny and Granddaddy, and Granddaddy hurried to tell Nanny all about his day and the game.

When I arrived, Nanny knew already that we had lost. She also heard that I had hogged the ball and failed in my attempts to score from the two-yard line. Not one to hide my feelings, my mood was foul. Coke and cookies helped a little, as did Nanny's hugs, but my pouty face gave away my sadness.

Granddaddy got out the poker chips and cards. "Wanna play a hand or two before supper?" he asked. Not waiting for an answer, he began to deal. "Michael, you had a rough game today. Let's talk about it," he added, even though he'd been there and congratulated me afterwards on a good game.

I told him how Tommy Wadden had scored, how the fourth graders made a two-point conversion, and how we were stopped at the end of the game at the goal line. I did not dwell on my glory-seeking; he knew all about that already.

"Sounds like you tried your hardest. Didn't work out, did it?"

"No," I replied, on the verge of tears, "it didn't." Then I fell silent, my chin quivering, not trusting myself to say anything else.

He went on, "How'd your teammates take it?"

It took me a while to respond. Why is he asking about my team-mates? A little confused, I replied, "OK, I guess. I don't really know. No one said much afterwards." I stared into my Coke glass.

"Michael, let me tell you something, something important. Look at me. I learned long ago that it is easy to second-guess people, easy to be on the sidelines and jump up and down. It is something else to be in the arena. Never forget that. It's important. I hope you understand how proud your grandmother and I are of you, for trying your best. It did not work out for you today. I bet you learned a valuable lesson. Next time, you never know. The most important thing is, never give up, OK?"

His words were cold comfort. I wanted the glory, to score a touchdown like my friend Tommy Wadden. I felt guilty that I wanted the glory and guilty that we had not won. That was the truth of it. Pride and vanity, two of the seven deadly sins.

Granddaddy pulled me next to him on the sofa, put his arm around my shoulders. He smelled of pipe tobacco, ash, and bourbon—as always. Some things never changed. He didn't say anything. He'd already shot that bolt. Just sat there with me and let me work it out in my mind, or try to work it out. I guess that was his point; it was my dilemma to work out. Most importantly, as I struggled to come to terms with my sadness and anger, he would stay by my side.

After a bit, Granddaddy got up to fetch a book about Theodore Roosevelt. "Listen to this," he said, and then he read:

> It is not the critic who counts; not the man who points out how the strong man stumbles, or where the doer of deeds could have done them better. The credit belongs to the man who is actually in the arena... if he fails, at least he fails while daring greatly, so that his place shall never be with those cold and timid souls who neither know victory nor defeat."

Teddy Roosevelt spoke those words in 1910 in Paris. The truth of them rings true today, and when I heard them for the first time as a child, I said to my grandfather, "But Granddaddy, we lost. I should have given the ball to Tommy Wadden."

"Well, yes, maybe so," he replied, "but you didn't, did you? What if he had fumbled the ball? Did you consider that? Lots of things might have happened. You made decisions that did not work out. You did not win. And I bet you'll never forget it, ever. But, you know, that's a good thing. From now on, you'll remember how much it hurt to lose, especially when winning was so close. And that'll help you to prepare hard, not just in sports, but in other aspects of your life, so it won't happen again. You'll see."

Granddaddy went on, "Let me tell you a story. I met Winston Churchill a couple of times. Once when that picture in the back hall was taken, and another time about nine months before. We were on the *Queen Mary,* crossing the Atlantic in September 1944. He had been meeting with President Roosevelt, one of many meetings between the two of them.

"I was invited to lunch with Churchill along with several other American general officers onboard. I suppose Churchill felt it to be his duty to meet with us, to entertain his American allies. Regardless, I was thrilled. He was a titanic figure, the voice of reason and determination against Hitler and the Nazis. I knew full well that he had stood alone in 1939 against the rising threat of German militarism, not only in his own country, but around the English-speaking world. I admired his clear thinking and his speeches—speeches that rallied both the English and Americans to the cause of freedom.

"I was the senior American general among our group. We gathered in my stateroom before lunch. Everyone was nervous, worried about the upcoming meeting. They wanted to be good guests but had so little in common with Churchill, a British aristocrat, that they fretted that conversation would lag. I poured everyone a little snoot of bourbon, a little liquid courage. It helped. Grim facial expressions relaxed a bit; then it was time. Commander Thompson, Churchill's private bodyguard, knocked and entered. He took his duties as bodyguard seriously. He was unapologetic about the need to search each of us before proceeding further. Professionally, and I must say thoroughly, he gave a quick pat-down, and then we were on our way."

"Granddaddy," I asked, "why are you telling me this story?"

"Michael, the point is Winston Churchill suffered many defeats in his life. After each, he got back up and got back into the fight. He often said, 'Never, never, never, never, never, never, never give up.' Seven nevers. If he can do it, so can you."

Although Granddaddy misquoted Mr. Churchill, the gist of it was true. Here's what Mr. Churchill really said, "Never give in, never give in, never, never, never, never—in nothing, great or small, large or petty—never give in except to convictions of honor and good sense. Never yield to force; never yield to the apparently overwhelming might of the enemy."

I wiped my nose.

Granddaddy then asked, "You ready for dinner?"

I nodded, feeling a little better, though not much. I needed to get something to eat. In retrospect I realize that Granddaddy knew this football game for third and fourth graders was no different than leading men in battle; there were always winners and losers, victors and casualties. How the game was played was most important thing. Mistakes were opportunities to learn. To learn, it was important to talk it out, without fear or recrimination. Then it would be time to eat and move on. He was teaching "his little soldier" a lesson from one early skirmish in the great war of life. And then the lesson was over. What I learned was up to me. He'd done his job.

Interestingly, years later I asked my brother Scott about his meeting Generals McAuliffe and Spaatz. What Scott remembers most about that brief encounter was General McAuliffe's face and his penetrating blue eyes. He was a handsome man, square-jawed, fit and athletic, befitting a retired paratrooper. He recalls nothing about the game.

But I do. Boyhood stories from a time not too long ago hold lessons learned on the playing fields of Landon. Like those learned on the playing fields of Eton, they developed character and helped us win future battles, whether for civil rights, equality, or victories over enemies foreign and domestic. I like to think that Mr. Banfield would be pleased. His vision of a school dedicated to character development was real and true. Banfield believed the principles of fair play, discipline, and self-restraint were timeless. We would quote Rudyard Kipling, "If you can meet Triumph and Disaster, and treat those two impostors just the same…Yours is the Earth and everything that's in it, And —which is more—you'll be a Man, my son."

Recently, one of my Landon School classmates from the Class of 1971, Sandy Gordon, reminded me of Banfield's basic decency. Gordon struggled academically during his years at Landon. He was also always cheerful and engaging. Banfield made sure Gordon received the extra attention he sometimes needed in the classroom. When Gordon came up to shake Banfield's hand at commencement in June 1971, Banfield pulled him in close and whispered, "We made it. Congratulations!" A moment of warmth and decency Gordon has never forgotten.

In later years, when I read Sir Matt Busby about playing soccer in the right spirit, I was reminded of Banfield's insistence on team sports and comradeship. Rules governing the outcome of hard-fought contests maintain honor, whether one wins or loses. So long as one plays to the fullest of his strength and endurance, dignity is preserved.

The Army–Navy game is another athletic event dedicated to these ideals. Tommy Wadden and I often attended the game as kids; believe it or not, our parents would put us 11- or 12-year-old boys on the bus at Westmoreland Circle with our game tickets and a few dollars. In fact, one year we had no tickets, just instructions on how to buy game tickets in the parking lot from scalpers. That's how we witnessed Heisman Trophy winner Roger Staubach play his last game in a Navy uniform, losing to Army, in November 1964.

In later years, my father would take me and my family to the game. He was particularly interested in my daughter Emma's soccer career, thinking he might find in her a family member to follow him to the Academy. It was not to be, but his re-introduction of the game led me back. In fact, Tommy Wadden (now Dr. Thomas A. Wadden of the University of Pennsylvania) and I find ourselves standing at the end of the game singing "Navy Blue and Gold" and "Hail Alma Mater," the West Point Alma Mater, with tears in our eyes, me especially.

7

Nanny

When the Landon School for Boys was founded by Paul Landon Banfield in the summer of 1929, he worked with a vengeance to transform his dream into reality, beginning by renting a mansion at 21st Street and Massachusetts Avenue. It was a heady time, the Roaring Twenties. With hard work, dedication, and vision, anything seemed possible, not only for Banfield working to start his school, but also for my grandparents, ever alert to the perils and opportunities of Army life.

Nanny played a significant role in Granddaddy's many successes. She was born in 1901 in Columbus, Georgia, a small town in the deep South on the Chattahoochee River, along the Georgia-Alabama border. One of two children and the only daughter, she was adored by her father, George C. Palmer, a lawyer. She was a quintessential daddy's girl and they shared a special bond. He especially admired her intelligence and willfulness, evidenced by an essay the twelve-year-old wrote in support of the suffragette movement:

> If the negro and illiterate white man is allowed to vote, why not an intelligent woman? Woman has to pay taxes as well as man; and she should have a say-so about what is to be done with her money. All women that believe in woman's rights are not militants. The suffragette in England would not break windows and set fire to buildings, if the people would

give her the vote. Woman has asked for the right to vote in the right way; but she did not get it. So, she is going to show the world that she will have it.

Woman is as strong mentally as man. There is no doubt that in a few years women will rule the world. If a woman has a husband or a father to vote for her, it is all right; but when she is left without either, she must take care of herself. MEN HAVE RULED THE WORLD LONG ENOUGH. WOMEN SHOULD HAVE SOME AUTHORITY NOW.

The Constitution of the United States of America says:

"The right of the citizens of the United States to vote shall not be denied or abridged by the United States or any State, on account of race, color or previous condition of servitude."

Thus, it will be seen that the right to vote is given citizens. The question that arises, is, who are citizens? The same Constitution defines citizenship to be as follows:

"All persons born or naturalized in the United States, and subject to the jurisdiction thereof, are citizens of the United States, and of the State wherein they reside."

By the above, it will be seen that a woman has the constitutional right to vote. Why not give it to her?"

Convincing words from a twelve-year-old, though I suspect George C. Palmer, the lawyer, had a hand in the writing. Even so, Nanny's essay shows confidence and willfulness, traits confirmed in her high school yearbook description: "Sue is our champion; she gets the dog for chewing the 'rag' as well as gum. She observes 'banker's hours,' and we seldom see her before 9:30."

My grandparents met in the autumn of 1923 in Columbus, Georgia. Granddaddy was stationed at the Infantry School at Fort Benning, in command of an artillery unit. Sue was a comely, vivacious brunette, spoiled by her father, whose mother treated her to New York City shopping trips twice a year. When they met at a tea dance at the Officer's Club, it was love at first sight. Though a confirmed bachelor after his divorce from his first wife, Grace, Granddaddy didn't stand a chance.

Nanny's mother, as strong-willed as her daughter, just could not understand her daughter's infatuation with this man ten years her senior. She wanted her daughter to marry a good Southern boy from a nice family, raise her children in Columbus, and live out her years near her! Sue would have her way. Coming of age in the Roaring Twenties, with the right to vote and stories of the Civil War fading with each passing year, Sue wanted to see the world. The fact that Granddaddy had been married once before may have troubled her mother, but apparently not her.

Nanny saw the world, following my Granddaddy all across the country and halfway around the world. From Columbus, they were transferred to Fort Leavenworth, Kansas, to the Command and General Staff College. While Nanny cared for her newborn daughter, Sue, or "little Tudy," Granddaddy studied diligently and was an honor graduate (top 10%) of the class. From Kansas, it was off to the 24th Field Artillery at Fort Stotsenburg in the Philippines. The tropical heat reminded Nanny of Georgia; two years later, in 1927, they transferred from Fort Stotsenburg to the Army War College in Washington, D.C., where she found the spring and autumn weather delightful.

She also found Washington to be an exciting town. Things were happening; the people were interesting. When Granddaddy was posted to the War Department, he worked in the State, War, and Navy Building on 17th Street, an elegant Beaux Arts building that housed the State Department, as well as the War Department and Navy Department, in the days before the Pentagon. Granddaddy worked shoulder to shoulder with the best and the brightest—not just military officers like George Marshall, Dwight Eisenhower, and William Simpson, but diplomats and politicians, the most important of whom was Secretary of State Henry L. Stimson.

Stimson already had a long record of public service in 1929. From 1911–1913, he served as Secretary of War under President William H. Taft, a fellow Yalie. When war broke out in Europe in August 1914, Stimson joined General Leonard Wood, ex-president

Teddy Roosevelt, and Elihu Root as one of the original members of the Preparedness Movement. They believed in a "realistic" approach to world affairs, based on strength and military muscle, not idealistic crusades, such as democracy for all or national self-determination. The Preparedness Movement evolved into the Plattsburg Movement, a series of summer camps in 1915 and 1916 where 40,000 men with anglophile leanings prepared for war. Anxious to do his duty in the fight against the Kaiser, at age 50, Stimson joined the National Guard, trained at Plattsburgh, deployed to France, and attained the rank of colonel in the National Army.

Granddaddy had been introduced to Henry Stimson at the Army War College in Washington in 1927. Stimson was on his way to the Philippines as Governor-General under President Calvin Coolidge. Stimson picked the brain of the newly returned Major Anderson. They renewed acquaintances in Washington, comparing notes of their experiences in the Philippines. Since they both worked in the State, War, and Navy Building, they saw each other frequently and discovered they lived quite near each other in the leafy suburb of Cleveland Park.

Though they lived close to each other, their circumstances were vastly different. Stimson was East Coast aristocracy, born in Manhattan in 1867. He was the son of a surgeon, not the son of an Iowa farmer, but his life was not without hardship. When Stimson was nine, his mother died of kidney failure. He summered in the Catskill Mountains of New York and attended Phillips Academy in Andover, Massachusetts. He graduated from Yale College in 1888, a member of Skull and Bones, a secret society that opened many doors in Stimson's future career. He finished Harvard Law School in two years and became a Wall Street lawyer. After making his fortune, he entered government service, first as a U.S. Attorney. With the many privileges of his station and class came a deep and enduring obligation to serve his country. He ran unsuccessfully for New York governor, followed by appointment as President Taft's Secretary of War. During World

War One, he became an artillery officer in the trenches in France, experiencing first-hand the horrors of trench warfare.

He was married to Mabel W. White, the great-great-grand-daughter of Roger Sherman, a signer of the Declaration of Independence. In 1929 the Stimsons bought the Federal-style Woodley Mansion, built in Washington, D.C. in 1801 by Philip Barton Key, an uncle of the composer of the "Star-Spangled Banner." It was home to ex-presidents Grover Cleveland, Martin Van Buren, and James Buchanan. The Stimsons were happy to show off their new home. They became famous in Washington for having small family gatherings as well as for hosting larger events, replete with famous military and political figures like retired General of the Armies John Pershing, and rising stars like General Douglas McArthur, and Majors George Patton, Dwight Eisenhower, and William H. Simpson. Years later, as a Landon School third grader, I would play soccer on the lawns of Woodley Mansion, then the home of the Maret School. Of course, I had no idea of the significance of the fancy buildings and grounds. Many important figures in U.S. history, including Major George S. Patton, Secretary of State Henry Stimson, President Grover and Mrs. Cleveland, Adolf Berle, Captain Hayne and Sally Ellis, General Lorenzo Thomas, and Senator Frances Newlands, lived, worked, rode, drank, conferred, and schemed within its walls and in its gardens.

The Stimsons had no children, Stimson having suffered a case of adult mumps that left him sterile. Mabel Stimson was a generous and kindly woman, often inviting Nanny and Granddaddy, and little Tudy, my mother, to visit. Mrs. Stimson took Nanny under her wing, seeing much of herself in the vivacious mother with an active daughter. Woodley Mansion was situated at a higher elevation than much of Washington, in the shadow of the Washington National Cathedral. It was an airy and comfortable estate, with a welcome breeze on those humid summer days before air-conditioning. With an expansive lawn where children could play, Woodley provided the

Andersons a welcome retreat from their cramped home on nearby Garfield Street.

Visits to Woodley were a mixture of business and pleasure. The Secretary of State could observe and evaluate the behavior and character of men and women who might be valuable to him. In the summer of 1929, Stimson needed someone with real wartime field experience to go to Geneva for a conference hosted by the International Red Cross. The assignment would entail both social and professional activities. The wives of the members of the delegation were expected to attend receptions, dinners, and galas put on by their hosts. He tapped my grandfather, and by extension, my grandmother, for the job.

The United States delegation to the International Red Cross Convention relative to the Treatment of Prisoners of War would be led by Eliot Wadsworth, the president of the American Red Cross. Granddaddy would be the only member of the United States delegation in Geneva with experience in the field in World War One. He had been with the French in the line, he had gone into "No-Man's Land" with the British, and he had been gassed in Picardy. He had seen Germans taken prisoner, both wounded and unscathed. He had seen some of his own men killed by Germans. He knew the rage felt by the survivors of combat.

Granddaddy was the perfect choice to be the Army's "technical advisor" to Mr. Wadsworth. Nanny's talents were no less important. Talkative and attractive, she had a working knowledge of French, and the coquettish manner of a Southern belle. Travel to Europe on the White Star Lines HMS *Majestic* thrilled Nanny. Seeing her name on the passenger list among the "Who's Who" of the diplomatic world confirmed to her that her decision to marry this son of Iowa, a farm boy with ambition and intelligence, was a wise one.

Nanny described the work of the Conference in letters home, describing how Granddaddy's long days in meetings were followed by social events that sometimes stretched into the wee hours. In one letter to her mother, she wrote, "On one side of me was a Turk with only rudimentary command of English. Didn't have much to say to him. On the other side was an Australian general, much more to my liking. We had a grand conversation."

"Fêtes" in Geneva were common, and day trips to glaciers and famous landmarks like the Château de Chillon, built on an oval limestone rock jutting into Lake Geneva (and the setting for Lord Byron's poem "The Prisoner of Chillon" and Henry James' novella *Daisy Miller*) helped pass the time. Nanny's "proper" education helped her appreciate her surroundings and endeared her to the Swiss hosts of the International Red Cross.

When the Convention ended later that summer, my grandparents' sojourn overseas continued a little longer. Although anxious to see their "little Tudy," who had stayed behind in Columbus, Georgia with Nanny's parents, Daddy George and "Mommee" Palmer, Nanny and Granddaddy traveled separately for a few days before embarking for home. Granddaddy wanted to see Koblenz, Germany, the confluence of the Rhine and Moselle rivers and the site where the Sixth Field Artillery had crossed into Germany in early 1919. Little did he know that twenty-five years later, he would fight his way across the Rhine, a short way downstream from Koblenz.

Meanwhile, Nanny traveled to Paris with her maid, Sarah, to shop for herself, her mother, and her four-year-old. Paris had fully recovered from The Great War, and luxury never seen in New York was available there. Nanny had enough money from her mother to load her steamer trunk with clothing and lingerie. Granddaddy joined her at the George V Hotel in Paris for a few days before crossing the English Channel for a visit to London to see the sights of Parliament, Big Ben, and Buckingham and Kensington palaces. Granddaddy ordered a pair of custom-made field boots from Thomas Bootery, 5 Saint James Street, a luxury befitting a rising star. Among his Army friends back in Washington, only Lt. Col. George Patton could afford, much less find, a finer pair. No longer would

Granddaddy be embarrassed to join Patton and Pershing on their early morning horseback rides along Rock Creek Park.

For this young couple at the height of the Roaring Twenties, the possibilities seemed limitless. My grandmother's dreams of getting the vote, of enjoying an exciting and fulfilling marriage, of raising children among high society, and of leaving the Deep South with its prejudices and sad history were all coming true. Their home in Washington, D.C. became a monument to their achievements, with medals, memorabilia, and photos that inspired me at the time, and which continue to serve as keys to opening up my understanding of my grandparents' strengths and weaknesses.

8

Sacrifices

When we are young and everything is going our way, it can seem as though the sky's the limit, that anything is possible. So it was with my maternal grandparents, the young married couple in the 1920s who came of age in the farmlands of Iowa and the heat of southern Georgia. Now they were raising their daughter, little Tudy, among the Washington elites, being noticed in the halls of power, and entrusted with delicate diplomatic assignments. Their lives were evolving in ways they could hardly have imagined.

Then the roof caved in.

The stock market crash of 1929 sent the country into a financial tailspin, a crisis deeper and more traumatic than any the country had ever seen before. Frantic to improve the situation, President Franklin D. Roosevelt decided government spending should be decreased. Granddaddy and all Army officers took a 15 percent pay cut. Their pay, albeit smaller than desired, was at least steady and guaranteed. Roosevelt also froze all Army officer promotions. Granddaddy, who had reverted to his regular Army rank of major in 1920, remained a major until 1935.

The citizens of Columbus, Georgia were proud of their soldiers and happy for their presence in their small city. The Infantry School at Fort Benning meant officers and their families were constantly in and out of town buying the necessities the Army did not provide. The local merchants flourished. How fortunate for the citizens of Columbus that Nanny's father, my great-grandfather George Currell Palmer, helped convince the Federal government to establish the Infantry School at Fort Benning in Columbus in 1917.

At the time, it wasn't an easy sell. Opposition to the Georgian's proposal to locate the Infantry School at Fort Benning came from an Ohio delegation who wanted to locate the Infantry School at Camp Perry, on the shores of Lake Erie. The Ohioans argued the summer heat in Georgia was too severe for Army training. In response, the Georgians argued that soldiers needed to be prepared for battle anywhere in the world. The Georgia summer heat and humidity mimicked hotspots like the Philippines, Cuba, and North Africa. The Georgia winter was mild enough that the Army could train year-round. Most importantly, the Georgia climate meant the Army could forgo the expense of building and heating structures capable of withstanding Ohio winters. Good points, all.

The Secretary of War, Newton Baker, wavered. Although the Georgian's economic argument was sound, Camp Perry was only 80 miles west of Cleveland. The one-time mayor of the city was hard-pressed by his fellow Ohioans. They argued he owed his former constituents the benefit of another federal installation near the city. But Secretary Baker had other, more pressing concerns. The desires of Cleveland were secondary to the needs of a nation in a desperate war in Europe. "We need to use all our resources wisely. If the Army can train soldiers more cheaply in the South than in Ohio, that's what we ought to do," he reasoned.

Baker weighed another factor, the pernicious memories of the Civil War. He believed they clouded too many decisions originating in Washington. He believed it was time to let bygones be bygones, to get past the corrosive notion of "Yankees" and "Rebels." The choice of Columbus, Georgia and Fort Benning over Cleveland, Ohio and Camp Perry would demonstrate high-mindedness, a reasoned and practical solution. Baker's boss, President Woodrow Wilson, who prided himself on an intellectual approach to delicate issues, would be pleased that facts, not prejudice, carried the day. And the decision would generate good will in the Deep South, no small matter for someone like Baker, who harbored hopes of someday running for president himself.

In the end, a few rounds of Southern whiskey and a bit of Southern charm sealed the deal. The Infantry School was going to Fort Benning in Columbus, Georgia. It was time to let bygones be bygones, or so it seemed.

Columbus was a Southern town with deep ties to the Confederacy. While the economic life of the city depended on the steady supply of federal monies, it would be unrealistic to think that those monies erased the bitter feelings many in Columbus felt toward Yankees about the losses they experienced in the Civil War. Many families had fought under the command of Generals Robert E. Lee, Stonewall Jackson, and John Bell Hood, flying the Stars and Bars. By the end of the Civil War, 92% of eligible Southern white males were in uniform. Many had lost loved ones in the fight. Half of the families below the Mason-Dixon line had lost either a son, brother, or father. They also lost innocence, treasure, and a way of life—rightly or wrongly, the losses were terrible.

From the perspective of 150 years later, one can say the South got what it deserved and that the pain and suffering of enslavers who lost everything in the rebellion against the North was justified. One can say harsh penalties are an appropriate remedy for the great evil. However, my study of the times and my family reveals a more complicated story with plenty of blame all around. For example, the government of the United States called on the South for men and material to support the fight against the British in the War of 1812. In fact, the bales of cotton in General Andrew Jackson's ramparts that stopped British musket balls at the Battle of New Orleans were the product of enslaved labor.

When the United States annexed Texas in 1845, Tennessee Volunteers were called to "Remember the Alamo!" and fight against Mexican General Santa Ana in the Mexican War. The wool that kept the soldiers of the United States Army warm during cold winter nights was produced and harvested by enslaved labor. Whether it was Northern merchants who traded in Southern agricultural goods or the English merchants who bought its cotton, no one was blind to the economic system that exploited enslaved labor. For example, despite President Abraham Lincoln's 1863 Emancipation Proclamation, the state of New Jersey delayed manumission until January 23, 1866. White Southerners were not the only ones reluctant to stop enslaving Black people. In fact, too many in both North and South were party to the horrors of slavery.

The Palmers were one such family. In 1932 George Palmer's sister Martha, my grandmother's Aunt Mattie, prepared an article for the Lizzy Rutherford chapter, United Daughters of the Confederacy. A spinster, and a librarian in Columbus, she had it published in *The Columbus News*. Aunt Mattie's father, John Palmer, was Nanny's grandfather, or my great-great-grandfather. While the memory of the "Lost Cause" was very real in 1932, I believe the article is about reconciliation.

> Sam and I [John Palmer] started off on a trip, a sort of march of triumph, where he would have everything our own way, which meant, a good easy time, lots of glory, and the stars and bars forever in the ascendancy. We did not travel as we planned; for we changed our route several times; but that's nothing, for tourists frequently do that, you know. Sometimes it is their fault; then again, it may depend upon circumstances.
>
> Uncle Sam, a veteran of the Mexican War, with the chase they gave Santa Anna, no doubt, still uppermost in his mind, took us aside one day, and, with his usual dignified manner, said very seriously and impressively: "Boys, your father being away from home and I am leaving for the East, I believe it to be my duty to give you some advice. War is terrible, and

don't you enlist; you are under the age limit, and they can't make you serve, so stay out of the army."

We promptly obeyed him by running away from home and joining the first company that would take us. For the next year or so, we traveled again and fought, too; this time, men who spoke good English, me, who spoke poor English and men who spoke no English at all; foreigners who did not know nor care what they fought about so they were paid so much, per day.

One morning, a red-faced Irishman big and broad, who towered above me, in all the impertinence of his six feet, laid a huge hand on my shoulder, and, in a deep bass voice, said: "You are my prisoner." I was only a stripling, lacking four years of being able to vote, and had no comrades except some in the same fix, I meekly surrendered. The meekness was all on the outside, though for, inwardly, I was as mad as a hornet. I faced the situation as bravely as I could, and with the other boys, marched through Knoxville to the county jail, which had been converted into a military prison.

On the way, whom should I meet but Sue Boyd. I never once glanced in her direction. She and I had been friends once; we had even gone further than that: I had sent her a valentine that Sam had written for me, he being the literary one; it was in three verses, "To Brown Eye'd Susan." We lived just opposite to them, as far back as I could remember, they had occupied the old Governor Blount residence. Our grandmothers had been reared together like sisters; our mothers were schoolmates and friends, and we boys and girls were as one family. They were Union people, strongest kind. Sis and the girls fell out first; then, we boys took it up, until both families quit speaking altogether.

When war was declared, Father was in Texas buying wool, so as to corner the blanket market; but it turned out that he was wool-gathering; for no sooner had he bought it, then it was confiscated by the Confederate Government. His friends interested themselves in the matter and it was recovered. Then, it fell into the hands of the Yankees, and he never saw that wool again. After the war however, the Federal Government paid him half of what it was worth. They came very near, catching him along with the wool; but he managed somehow to slip through the lines, and, because of that fact, Sue Boyd remarked that "he was no gentleman."

I stayed in Knoxville jail for some time, and Mother and Mattie, my cousin, [not the author of this article] used to visit me whenever they were allowed the privilege. Mattie's father was in the Union Army; but her two brothers and her sweetheart were with us.

It was against the rule for visitors to whisper to the prisoners; but Mattie never failed to do so when she came. Just as soon as she would begin, the guard would say, "No whispering to the prisoners." When she persisted as she always did, he would run his sword between us, then she would lean over the sword, and continue the conversation. He wasn't very fierce about it; for he knew that she did it just to tease him and not to communicate any Confederate secrets, or plan an escape.

One muggy, drizzly day, I stood at the window and stared out through the bars into the muddy below. I was feeling pretty blue; for I knew that I need not expect visitors in that weather, when who should come along but Phoebe, she who had belonged to the family before I was born. When she spied me behind the bars, you should have heard that ni**er holler. "Johnnie! Johnnie! I never thought to see you in that place," and then she commenced to cry. I turned and left the window, I was crying too and I did not want her to see me in tears.

In a few days, they brought Sam in, and then we went on another journey, this time to the military prison at Camp Douglas, near Chicago, where we tarried again for some time.

At first, visitors were allowed in the camp on certain days, and those of us who were handy, made things and sold them to the visitors. With my pocket-knife, I made the spending money for both Sam and myself, carving baskets from peach seeds, photograph frames from cigar boxes, and crochet needles from old bones.

Captain Sponnible stood guard at one of the gates; he had more scars on him than I ever saw on any one before, and had been shot so many times in the legs that he looked knock-kneed when he walked. He was too crippled for active duty, yet game enough to act as guard.

One day, as several of us stood near him, two women came in, looked us over, and one sneeringly said, "Are they soldiers, those dirty things?" He fairly swelled, he was so angry, "By gad! Madam, yes, these men are soldiers. They've been fighting, you take those things up there," and pointed to some soldiers who had never done anything but guard duty, "let them do some fighting, and they'll be ragged and dirty too."

At last, the time came when we were told that the war was over and we could go home. The prison door swung open to let us out, the gate closed behind us as we walked into the street. We were free as the birds of the air, exulting in the blue sky above us and the green turf beneath our feet, yet not a spot of ground nor a tiny corner under even the humblest to call our own.

Our clothes were so tattered and torn that even the ragman might have passed us by. We had on Confederate gray pants made by a Tennessee lady whom we had never seen, and they were voluminous as to width of leg; we wore the long-tailed Union coats, with the tails cut off even with the waist, so that we would not be mistaken for Union soldiers and escape; our shoes were several sizes too large for us, and all in all, we looked like two old scare-crows left out in the field to be blown about at will by the winter winds. We had not heard from the family for more than a year and did not know what changes had taken place in the meanwhile; we had very little money, for the camp had been closed to visitors for some time, and we did not know how to make our way back South.

Sam and I stood on a street corner and talked over the situation for some time. We finally decided that it would be best for us to go home separately and when we arrived at this conclusion, we lost no time, bade each other good-by, and I started to journey again, this time alone.

I walked down the street of a small Western town a few days afterwards. It was then about eleven o'clock, and I had no breakfast, when I chanced to look up and saw tacked on the side of a store, a sign, "Painter wanted." I knew as much about painting as I did about "heathen Chinee" but, I was hungry, and I made up my mind that I would have that job. The proprietor listened to me, pushed his spectacles up until they perched like a parapet on top of his bald head, looked me over, and then said: "How do you mix your paint?" This almost made me lose my breath for a moment; but I shot back at him: "Tell me how you mix yours and I will tell you how I mix mine." He explained very carefully how he mixed his paint, and when he finished, I told him, "that was just the way I mixed mine."

I was given the job and so thoroughly did we mix paint alike that I stayed with him several months, as painter, clerking in the store, and did the chores around his home between times.

A letter, with check enclosed, came to me one day from Uncle telling me to join my brother at a certain place, and together, go to a small town in Georgia, where the family had refugeed a year previously.

To tell the merchant and his wife good-by was a hard thing to do. They had taken me, a stranger, into their humble home, and we had become deeply attached to each other. In as firm a voice as I could, I told them of my letter, my family, my homesickness, and with a lump in my throat, packed my trunk and started on my journey again.

When I joined Sam, we traveled as far as we could on the railroad, then went the rest of the way by private conveyance.

The wheels of the rockway ploughed deep sandy ruts through the long stretch of road that led to the village, where eleven miles away, a few months before Jefferson Davis was captured, not in woman's attire, but clad in the habiliments of the man and gentleman he was. Gone were the hopes and dreams of the Confederacy! Gone! Everything gone! Save the principles for which we fought, save home and family. And home meant the little abode just beyond us at the turn of the road, so close to us that, even now, we could see the homemade, homespun, ruffled white curtains as they flapped a welcome back and forth to us through the tiny windows of the times-seasoned and weather-beaten log house.

Mean and insignificant as it seemed, it was home and home meant the family. They were all there but one, Frank, the baby. The long tedious trip through a part of Tennessee, West Virginia, North Carolina, South Carolina, and Georgia; the change of climate, and the privations undergone were more than he could stand. In the church-yard near, the pines crooned a lullaby, the mocking birds sang their sweetest in among the waving boughs over-head, and little Frank calmly and peacefully slept through it all.

Into the house, with its unchinked walls, we went; back to the Mother who thought she had given us up for the honor and glory of the Cause long ago; into her arms first and then into the arms of the others. No good, easy times, nothing our own way, the stars and bars trailing mournfully in the dust; two weary, dust-laden, heavy-hearted Confederate soldier boys knowing nothing of war, and military tactics, had fought as good a fight as they could. In the year of our Lord, 1865, the journey ended.

John Palmer's brother Sam was an artist, good enough that his sketches were included in American Heritage's *Civil War* published in 1961. The book was a large, coffee-table tome, full of vivid pictures and descriptions of the Civil War and marketed to capitalize on the 100th anniversary of the war's start in 1961. I remember spending hours poring over the maps and photos, an eight-year-old boy enthralled by the romance of it all. I was not the only one; Nanny said, "Those sketches look exactly like Uncle Sam's work." And so, they were, similar in style and content to sketches in Sam's diary that Nanny owned.

Nanny contacted American Heritage and discovered the owner of the pictures lived in Washington, the curator of the Corcoran Gallery, Mr. Hermann W. Williams, Jr. In fact, Williams was opening a Civil War art exhibit, to commemorate the centennial of the start of the war. To be included in the exhibit were Sam Palmer's drawings. Nanny wanted to know how her uncle's work had come into Mr. Williams possession and discovered that Mr. Williams' grandfather was Major Samuel King Williams, a Union officer from Boston. During active service, Major Williams had been badly injured by a horse rolling over his leg. Unfit for active service, he was assigned to the Invalid Corps at Camp Douglas, a prisoner of war camp on Lake Michigan, near present-day Southside Chicago. There, Williams befriended many of the prisoners, including artist Sam Palmer. Major Williams exchanged food and clothing for Palmer's pencil sketches. Their friendship survived the war, revealed in letters between the one-time prisoner Sam Palmer and his one-time warden Major Williams:

Macon, Geo
Nov 7th, 1868

Dear Major,
    You will always be "Major" to me—I can't think of you as anything else. I don't know if this letter will ever reach you, but thinking of "lang syne" I resolved to write and see if you were still in the flesh—Well, I suppose you would like to know what I have been doing since I left the hospitalities of Camp Douglas. As soon as the war was over, I wended my way from Chicago a thousand miles south into the "piney woods" of Georgia, where I rejoined my family. That you may imagine was a joyful meeting. I went then to Columbus, Geo. and tried my hand at merchandise but failed in two years because of the hard times, paid all my debts though and came out clean-handed. In the meantime I met my fate in a Georgia girl and not having the fear of starvation before my eyes I married her some seven months ago. I have not yet regretted the step. I think I have the nicest little wife in the world (except yours, of course) and am as happy a fellow as the State contains.
    I suppose you will not be surprised to learn that, in politics I have been a supporter of reconstruction—I had the satisfaction a few days ago of voting for Gen. Grant, and I can assure you that is a rough road to travel in "Dixie"—However a better future is opening for the South. I look to see her prosper as she never did under the blight of slavery.
    I have told my mother about all your kindnesses to me

85

in prison, and I am sure she would be delighted to see you. Although she has a lingering dislike for "Yankees" (as most of the petticoat "persuasion" in this region have) yet she exempts *you* from the general condemnation. I differ so with all my family on politics except my wife, who like a sensible little woman does not meddle with it.

I often think of you and Camp Douglas and all the old set there, and I don't know what would give me more pleasure than to meet you once more. Remember me kindly to your wife and the boys and to the baby that was such a sunbeam to me at Camp. How I loved to look at it—just as I would at a flower growing in a cell, bringing refreshing memories of the sunlight outside—I suppose that is not the baby now. . . . I hope you have long since dispensed with your crutch. Please write to me and tell me all about yourself since we parted at Chicago—I don't like to lose my friends.

Alas, Sam Palmer and Major Williams were never to meet again. But the strength and spirit of their friendship endured. When Nanny and Major Williams' grandson Mr. Hermann Williams met, Mr. Williams was quoted in the *Washington Post* as saying, ". . . it was like talking to an old friend."

And so it was that Nanny's father, Daddy George, a Southerner, embraced a son of Iowa and a Yankee, Granddaddy, as his own. Granddaddy's Danish parents, the Andersens, were born in the 1840s and had died many years before, leaving him an orphan before age 30. Daddy George became something of a surrogate father, writing near-daily letters to Granddaddy, full of encouragement, jokes, and photos. While Nanny and Granddaddy were overseas in the Philippines in 1925, Daddy George sent news of Nanny's nephews, George and Martin, calling them "splendid specimens of childhood." Granddaddy returned the affection, writing Daddy George from Germany in 1929, "the beauty of the trip up the Rhine this morning begs description. . .I hope someday to repeat it with all of you." It was a nice sentiment, but one not to be; the rise of German militarism in the early 1930s ended any hopes of a return visit.

Regular Army officers like Granddaddy were among the first to recognize the rise of German and Japanese militarism, even back into the late 1920s. Granddaddy wrote in a letter to Daddy George:

> ...Japan, you will remember, succeeded in putting across a clause in the Washington Conference agreements that no additional fortifications or armament would be placed on any islands in the western Pacific. In other words, we lost out, and Japan got just what she wanted, for she had already fortified everything she wanted to fit to the full limit. Furthermore, Japan is probably violating the agreement if she sees fit to do so, or if she feels it to be to her advantage to do so, while we are religiously living up to it.

In 1932 the Japanese invaded and occupied Manchuria. In response, Granddaddy's mentor, Secretary of State Henry L. Stimson, instituted the "Stimson Doctrine" of nonrecognition of international territorial changes that are executed by force. Unfortunately for the United States military, President Roosevelt's election substituted appeasement for negotiations based on military strength. Despite all the evidence of Bolshevik predations and policies of mass starvation, the Roosevelt administration thought it best to recognize the Union of Soviet Socialist Republics (USSR) and to grant it most-favored nation status. The administration hoped to open the USSR for trade with the United States as a means of ending the Great Depression. Unfortunately, the administration also encouraged naïve Americans to move to the USSR in what turned out to be the largest mass emigration in the history of the United States.

The emigrants were promised jobs running automobile and watch-making plants that had been purchased and transported to the Soviet Union. Upon arrival, the American emigrants turned over their passports to Bolshevik authorities. Their citizenship renounced, they were considered Soviet citizens and often simply disappeared. Searching for alternatives, unwitting American citizens had believed the glowing accounts of the workers' paradise, of new Russian factories surrounded by trees, equipped with cafeterias, clinics, swimming pools, and nurseries. American Ambassador William C. Bullitt and his staff were powerless to assist in the repatriation of American citizens, many of whom simply vanished. Once an outspoken liberal, even a radical, Bullitt became openly hostile to the Soviet government and outspoken anti-communist.

Japan also began to ignore the norms of diplomatic practice. For example, the ship-building limits imposed by the 1936 Second London Naval Treaty were simply unacceptable to the Japanese, who began to build larger, more modern aircraft carriers and battleships.

Throughout these distressing times of international upheaval and national economic decline, Granddaddy was teaching at Fort Leavenworth, preparing regular Army officers, the small cadre of West Point graduates and World War One veterans, to be the next generation of staff and senior leaders. Officers like Carl "Tooey" Spaatz, Elwood Quesada, Matthew Ridgway, and Anthony McAuliffe were among his students.

During this same time, Daddy George's practice in Columbus increased to the point that he needed an associate. At a time when anti-Semitism was rampant in the Deep South, he chose the best young associate he could find, Aaron Cohn, a young Jewish lawyer. Just as Daddy George had seen something of himself in Granddaddy, he also saw something of himself in Aaron Cohn. Both of them had overcome great odds to become attorneys. In 1863 Daddy George's family had been "refugeed into the rebel lines" by order of the U.S. Provost Marshal of Knoxville, stripped by the Union occupiers of their property and Constitutional rights because they were deemed traitors to the United States. From Aunt Mattie's writing and Sam Palmer's letters we know how difficult times were in the South after the Civil War for everyone. Reconstruction was envisioned by Abraham Lincoln to reconcile the past with a new economic order without slavery. He envisioned the South returning to the fold as the prodigal son returning to the family. He envisioned a time of binding up the wounds of war without rancor or revenge. On the evening of the surrender of the Confederacy, April 10, 1865, Lincoln entertained a crowd and listened to bands at the White House. He thought "Dixie" a wonderful tune, too long usurped by the rebels, and is quoted to have said, "I have always thought 'Dixie' one of the best tunes I have ever heard. Our adversaries over the way attempted to appropriate it, but I insisted yesterday that we fairly captured it ... I now request the band to favor me with its performance." This is a concrete example of Lincoln's vision of magnanimity.

Unfortunately, Lincoln's assassination heralded a period of hate and revenge. Carpetbaggers and Republican politicians descended upon the South. While formerly enslaved people were now free, economic times remained difficult. Voting rights for Black men were curtailed by Jim Crow laws. The racial prejudices of the old South were embodied by members of the Ku Klux Klan, who viewed themselves as avenging knights.

The Palmers struggled to honor their ancestors while at the same time regaining some stability in their lives. Even today, we have opportunities to honor our ancestors in meaningful ways, respecting their sacrifices and honoring their achievements. To that end, my brother Scott and I recently donated the Civil War diaries and sketchbooks of Samuel Bell Palmer to the East Tennessee Historical Society McClung Collection. As I leafed through the pages one last time, I was struck by a limerick from the poet Lord Byron:

> A prodigal son, and a maid undone,
> And a widow re-wedded within the year;
> And a worldly monk, and a pregnant nun,
> Are things which every day appear.

Not only is the 18-year-old Sam Palmer languishing in the hellhole of a prisoner of war camp on the banks of Lake Michigan, he is recording from memory lines from *Manfred*, Byron's farce written in 1817. Perhaps Byron's skepticism of the Catholic church and its professed mores resonated with Palmer. And like all prisoners and young adults, Palmer's interest in the "fairer sex" shines through loud and clear. One of his sketches included in this memoir, "Italian Holliday," took the soldier to another world, where revelry and dance supplanted the day-to-day struggle to survive.

Attorney Aaron Cohn's family had also struggled to survive. Their journey began in the late 1800s, they faced the Czar's Cossacks, who murdered and pillaged the Jewish population of western Russia. The family survived and escaped to the United States, settling in Columbus, Georgia, in 1906. There, the small Jewish community was tolerated, though anti-Semitism remained and was still all too common.

Aaron Cohn was an accomplished horseman, the result of his father's livery business. He was an athletic man, an outstanding tennis player, and a fierce competitor. Never one to tolerate insults, Aaron Cohn thrived in the ROTC cavalry unit at the University of Georgia in the 1930s. Well aware of the history of Jewish persecution in Europe and elsewhere, but especially the increasingly prejudicial German laws, rising severe personal attacks, and ever more brazen atrocities in Germany, Russia, and Poland, Aaron Cohn prepared himself for war.

Daddy George knew war was coming, too, as he followed events overseas, knowing both his son-in-law and his new law clerk would soon be involved. When war was declared in December 1941, Aaron Cohn could have opted out of line service and taken an appointment as a judge advocate, the military equivalent of a lawyer. Instead, he sought service in a line unit, as a cavalry officer. His motivation and intelligence recognized, he became the S-3, Operations Officer, for the 3rd Cavalry Group, rising to the rank of major and leading troops of Patton's Third Army into southern Germany. Many days of harrowing combat and horror followed, culminating in the liberation of extermination camps in southern Germany. Cohn was undoubtedly aware that Patton's 2nd Cavalry had rescued the Lipizzaner mares out from under the advancing Soviet army in Czechoslova-kia and returned them to their home in Vienna. Author Mark Felton tells the remarkable story in his book *Ghost Riders*. Nearly 1,000 Lipizzaner mares from the Spanish Riding School, many in foal, were taken out of the grasp of the advancing Russian Army by coop-erating American and German soldiers.

Daddy George exchanged letters with Granddaddy in the spring of 1942, delighted to see that he was now Major General Anderson, the commanding general of the 102nd Infantry Division. Unfor-tunately, Daddy George suffered a series of heart attacks over the summer and died in September 1942, Nanny at her father's bedside. How he loved to follow Granddaddy's career! He would have marveled at the photograph of Granddaddy and Winston Churchill crossing the Rhine River in 1945.

Daddy George would not only have admired Anderson's career, but also Aaron Cohn's service record in World War Two, receiving the Silver Star medal for his bravery under enemy fire. Beyond that admiration, he would have been so proud of his one-time clerk's appointment as a juvenile court judge, and his many achievements during a 46-year career on the bench.

Daddy George, Nanny, and Granddaddy would delight to know my brother Scott and I met Judge Aaron Cohn on December 10, 2010, spending most of an afternoon with him. He called Daddy George, who became a Superior Court Judge late in life, "Judge Palmer" in this story of the two of them together:

"One day when we were walking down Eleventh Street to lunch, some of my 'cronies' yelled to us from the steps of the YMCA across the street, 'Hey, shyster!' Judge Palmer turned to me and asked, 'Do you know those ruffians?' 'No, sir,' I lied. 'I've never seen them before in my life.'"

Scott and I sat dumbfounded. What was Judge Cohn saying about our great-grandfather? Cohn chuckled, "If Judge Palmer had known my friends were making fun of his black-and-gray spats and long black coat with gray-striped trousers, I'd have been out of a job. I'm sure of it. I didn't like to lie, but it was a white lie."

We showed Judge Cohn the photograph of Granddaddy and Winston Churchill. It took a minute for him to realize that Granddaddy was Judge Palmer's son-in-law. After that fact sank in, I told him about General Eisenhower's order to Granddaddy to keep Churchill off the water and out of harm's way.

Judge Cohn said, "If a German '88' had hit that landing craft and killed Churchill, Alanbrooke, and Montgomery, the entire war would've changed."

He went on, "The British would have demanded revenge, used Churchill's death as leverage to get us into Berlin ahead of anyone else. We could've done it, too, a lot easier than the Russians. The Germans were giving up left and right in the west all through April. They didn't want to fight us; they would put up a little show of resistance and then quit. Not like earlier in the war. All they wanted was to get away from the Russians, who were slaughtering most prisoners, or sending them to slave labor camps. The Germans knew surrendering to us meant three meals and survival. Surrendering to the Russians meant death."

Judge Cohn added, "The Germans would've handed us the keys to the city. Eisenhower got cold feet. Asked his pal Bradley for a casualty estimate. Bradley said 100,000 dead. What nonsense! Your grandfather's XVI Corps took the northern half of the Ruhr Pocket, captured 300,000 Germans and lost 189 men. We had complete air and artillery superiority. We were just pounding them down, without having to resort to infantry attacks. But Eisenhower got the answer he wanted; he'd lost his drive. Eisenhower had begun to think like a politician, not a general. You think Grant would've stopped? No way. Eisenhower was irritated with the British. Found Montgomery nearly impossible and Churchill a spoiled brat. I think Eisenhower thought we no longer needed the Brits. We called the tune, so to speak.

"From what I learned later, Eisenhower wanted the war to end as soon as possible, for personal reasons. Eisenhower had an English-woman driver, Kay Summersby. They met in London, in 1942, before Eisenhower was famous. All the top American brass were given civilian drivers, usually attractive women who knew the streets of London. Because of the blackout, it was very easy to get lost in the twisty, narrow streets. Many of the street signs had been removed, as a precaution against German saboteurs, and in case of German invasion.

"Kay Summersby and Ike became quite the item. She followed him to North Africa, attended all the big conferences with him, and entertained him—pistol shooting, riding, and bridge playing. Back in London, they ended up in a small country cottage together. While there were other staff around, she made sure he got his rest, his cigarettes, and her attention. According to her book, *Past Forgetting*, they were deeply in love, the type of wartime romance that is powerful and consuming.

"Ike wanted to divorce Mamie Eisenhower, and his plan was to bring Kay to the United States as his aide, with the intention of marrying. Apparently, Ike discussed his plan with George Marshall, the Army Chief of Staff, his only boss other than the President. Marshall was appalled and told Ike he would do no such thing. Marshall added that should Ike divorce Mamie, Marshall would be sure that Ike would not succeed him as Army Chief of Staff."

Scott and I sat enthralled by Judge Cohn's description of events from more than 50 years ago. Here was a man, connected to the family via Daddy George and the pre-war Columbus days, who knew all about the ins and outs of Army politics. I felt like that six-year-old boy sitting with Granddaddy in the house on Albemarle Street, eager for more. In the basement of Judge Cohn's home were mementoes of his military service: medals, photos, and maps. Like Granddaddy's small display, or the wall of memorabilia Scott and I saw in 2011 in Judge Robert Morgenthau's office in New York City, the significance of the experiences represented more to Judge Cohn than we could easily understand—a time of service, sacrifice, and triumph. Against all odds, these men came through alive, and in their later years they contributed mightily to the greater good.

Judge Cohn began again. "So why do you think this Kay Summersby-Eisenhower romance is so important?"

My eyes widened and I looked to Scott. He shrugged his shoulders. We turned to Judge Cohn. He finished a drink of water. "Y'all want some?"

"No, sir," I replied. I wanted him to keep talking.

"OK," he went on, "Here's the real story. Eisenhower was so irritated that his romantic dream was blowing up in his face, and maybe a little embarrassed, who knows? I believe Eisenhower acted out in his decision to going outside normal channels, communicating directly with Joseph Stalin, to confirm the fact that the U.S. Army would stop at the Elbe River, leaving Berlin to the Russians. Instead of the President or George Marshall confirming the stop point decided upon months earlier at Yalta, Ike bypassed both of them, as well as the Combined Chiefs of Staff, the group that set all major policy decisions for the United States and Britain, subject to the approvals of President Roosevelt and Prime Minister Winston Churchill, and informed Stalin that the U.S. Army would stop at the Elbe River.

"In early February 1945, it was agreed that the Elbe would be the stopping point for Allied forces. At the time, the Allies were still recovering from the Battle of the Bulge. Even though the Allies were starting on the offensive, we were more than 200 miles away from Berlin, while the Russians were only 40 miles away. In early February, letting the Russians take Berlin seemed to make sense. Not now.

"We were pressing forward in 10- and 20-mile leaps against token German resistance. In the east, resistance was stronger than ever. The Russians were sacrificing thousands of soldiers for advances of hundreds of yards, still 40 miles away from the center of Berlin. German soldiers were fleeing to Allied lines, swimming across the Elbe in groups of hundreds. The prospect of capture by the Russians was terrifying.

"There was no good military reason to abandon the longtime and oft-repeated goal of 'On to Berlin!' It was ours to take and would have changed the map of Europe forever."

Scott asked, "Wasn't that a sound military decision? Spare American lives and let the Russians slug their way into Berlin?"

"Yes, it was, Scott," Judge Cohn replied. "From a military perspective, it was. But Eisenhower had no appreciation of the depth and scale of Stalin's treachery. Eisenhower's great accomplishment in World War Two was his ability to keep the Allied military coalition together. Keeping Montgomery and Patton from killing one another took every bit of his energy and skill. But with Stalin, Eisenhower was in over his head, as he would find out later."

Judge Cohn added, "In those days, there was a bright line drawn between the military and the civilian world. In America, civilians are in charge. In this case, the decision to stop at the Elbe was a political decision with huge political importance. Stopping at the Elbe meant abandoning much of Eastern Europe to the Russians. If the military facts on the ground had changed from early February 1945 to late March 1945, it was Eisenhower's responsibility to communicate those facts to George Marshall. George Marshall was responsible to the President.

"The President had long ignored the brutality of the Stalin regime. Stalin was ruthless in his persecution of the Ukrainians, and the Poles, and the Belarusians. Starvation with the death of millions was utilized in Ukraine to 'collectivize' the farmers in 1932. President Roosevelt ignored pleas for help. Up to that point, the United States had not recognized the legitimacy of Stalin's government. When Stalin offered to open Russian markets for American products, Roosevelt recognized the Soviet Union. The price for that recognition was turning a blind eye to the depredation of Ukrainians.

"President Roosevelt also turned a blind eye to Stalin's order to massacre 4,000 Polish officers, bury them in the Katyn Forest, and blame the Germans. The lie was a convenient one; 'Uncle Joe' [Stalin] would never do such a thing. Turns out, the rumors were true.

"Stalin was paranoid that the Allies and the Germans would come to some separate agreement, a truce that would keep the Russians out of Eastern Europe. When Stalin learned of talks between American and German commanders to surrender all German forces in the field in Italy, he was furious. The talks in Berne, Switzerland, in late March 1945, had reached a tentative agreement—an agreement to stop fighting that would have spared lives on both sides. Stalin did not care about that; he was obsessed with the fact that a surrender would have opened the door to Austria and the Balkans to the Allies. Therefore, Stalin demanded that Russian negotiators be part of the talks, the goal being to delay as long as possible any agreement, the time being used by the Russians to push ever deeper into Eastern Europe.

"Better even than Roosevelt, Stalin understood the use of military power to achieve political ends. Any concession on the part of the Allies was an opportunity to be exploited. At Yalta, Roosevelt wanted very much for Stalin to join the fight against Japan. Roosevelt was unsure the atomic bomb would be sufficiently powerful to force the Japanese to surrender. In return for the promise of Russian participation in the invasion of Japan, Roosevelt agreed to cede most of Germany, and Berlin, to the Russians.

"More than anything, since the German invasion of Poland was the main reason the British declared war, the British demanded free and fair elections in Poland, with the expectation that Russia would allow a true democratic government in Poland. Stalin agreed to this demand, knowing the Allies had insufficient interest or power to force the issue. Once the Allies stopped, he would do what he wanted.

"Stalin remembered the history of the Allies in 1919. After

World War One, as part of the Versailles Treaty, the Allies demanded free and fair elections in Russia. Once Germany signed the treaty, the Allies lost interest. Despite all the high-minded rhetoric, they walked away. Stalin led his Bolshevik army to power, and in the process, millions died. Stalin's ruthlessness was not understood in the West, except by a few, like Winston Churchill. The Russians called the shots at Yalta.

"Last thing I want to say, then I gotta rest," Judge Cohn said. "I know firsthand the things the Nazis did. I saw concentration camps with my own eyes. As a Jew, I witnessed the horror of the Third Reich's 'Final Solution.' The victors write the history. Now, you tell the story, the history. Do everything you can so this'll never happen again."

Since my visit to Judge Cohn, my days have been filled with writing these stories in *General in Command* and *To the Front*.

9

From Washington to Bastogne

Growing up in Washington, D.C. in the 1960s was both heavenly and hellish. In early 1962, the doors were open for boys of a certain age and class. Institutions like the Landon School provided access to Mrs. Shippen's Dancing School, where dance, etiquette, and proper social skills were taught. We learned how to ask a girl to dance, how to "cut in," and how to mind our manners. Young men and women were groomed to be "nice young people," polite, responsible, and well-behaved.

Parties at the Sulgrave Club, an exclusive enclave of the Washington crowd, and Anderson House, the headquarters of the Society of the Cincinnatus, whose members must be a direct descendent of General George Washington's Revolutionary War officer corps, were commonplace. Tennis at the St. Alban's courts, golf at Congressional Country Club, and Saturday night parties at the Chevy Chase Club with Washington insiders. *The New York Times* reported that Ethel Kennedy loved "Hang on Sloopy," played by my brother Scott's garage band, the Decadents, comprised of tie-and-jacketed boys from Landon School at the birthday party for McGeorge Bundy.

When my father gave me his broken-down 1930 Model A Ford

with the words, "Make it run and it's yours," I was on top of the world. And I eventually got it running. When he showed me the trick of keeping the distributor cap in my pocket, I knew no one could steal the car I had worked so hard to make mine.

We had it made. All we had to do was work hard in school, get good grades, and get into the right Ivy League college. We would meet a woman from one of the Seven Sisters colleges, fall in love, marry, and raise a family. We would delay self-gratification and stick to the plan. With hard work and discipline, all the goodies of Washington would come to us, some sooner, some later.

From early childhood, I was fixated with my parents' wedding album, amazed by the black-and-white glossy photographs of their 1947 wedding. I admired my mother's silk gown, off the shoulders, the smiling faces of young and old, and crisp creases of the military uniforms, both Army and Navy, high rank and low, especially the wings of gold of the Naval aviators. Of all the photos, I cherished the one of my mother and Granddaddy, at the head of the aisle, gathering their thoughts for the last time as father and daughter, before they process toward the waiting groom and groomsmen, maid of honor and attendants, and Chaplain Kellogg, gathered together at the altar.

The seminal moment is captured in a flash, representing the fulfillment of a father's life. Granddaddy was never to wear the uniform again, the "pink and greens" of the U.S. Army, with ribbons. Granddaddy is drawn to his full height, not more than 5'8" but enough to complement my mother's 5'4". Face relaxed, eyes forward, chin in; how many times had he stood at attention, awaiting the command, the order, the right time to proceed. And so, it was here. The king ready to bestow his daughter on the worthy knight. As the music commenced, my mother squeezed his arm. Reassured, she embarked down the aisle to marry a Navy man, a graduate of the U.S. Naval Academy, Class of 1943, wearing wings of gold.

How is it that my mother, Sue Moore Anderson, married a

Navy man? Good question, and on the surface, one might think it surprising. Had she not dated West Pointers throughout her teen years? Had she not continued to be pursued by Army officers, some of whom were men under Granddaddy's command, men to whose character Granddaddy could attest?

My father, Harper Elliott Van Ness, Jr., known to all as "Smiling Jack," was movie-star handsome, possessed of a broad and ready smile, and a Naval aviator. A careful man and a combat veteran himself. A Midwesterner like Granddaddy. And like Granddaddy, an orphan from an early age.

Apparently Nanny was not pleased with my mother's choice of husband. When Nanny learned of her daughter's engagement to this Navy man, she hired a private investigator. What were his people like? Who are they? Is his family "good people?" Was Harper a fit match for her daughter or was he simply "marrying up"?

Whatever information Nanny's private eye discovered is unknown. Whether it was insufficient to deter my mother—or maybe the results confirmed my mother's good judgment—is also unknown. Most likely, Nanny's efforts to break up the engagement were stifled by Granddaddy. He likely reminded her that when he proposed to her in 1923, he had been divorced for five years. Like Harper, when he proposed to her in 1923, he had no money, no family fortune, only a determination to succeed in the service of his country.

Granddaddy also likely reminded her that families can be complicated. Was not Nanny's brother Randolph divorced from his wife Isabel, mother of Nanny's two nephews, Buck and George? I think Granddaddy likely reined in Nanny's tendency to think the worst of people, to think that no one is good enough for her daughter, and expected her to make the best of the situation, to control her tendency to impose her will on her daughter.

As I study the wedding photos, I see nothing to suggest anything but a fairy-tale wedding, the next step in Nanny's and Granddaddy's charmed and magical family life. All around them that day were friends and family from far and wide. As Granddaddy looked down the aisle, Nanny was seated with her mother, now a Southern dowager, dressed formally and completely in black. Since the death of Daddy George fifteen years before, she was not seen in public except in black. The finest silks, the richest fabrics and furs, fit her notion of class and status.

As my mother looked down the aisle, she took in the smiles of her husband-to-be. She had no idea how hungover he was. His groomsmen were members of his squadron, VF-20, and fulfilled their collective duty to drink the groom under the table. Since the wedding was an evening ceremony, there had been some time to recover appearances, but even for these young warriors, appearances were deceiving. All of them were feeling a little worse for wear.

On the other hand, the maids of honor looked and felt great. No late-night carousing for them. Like the maidens of Gilbert and Sullivan's *The Mikado*, the maids were all in a row, with lily of the valley bouquets providing a blissful scent to the packed church, covering the smell of wet wool and mothballs rising from the uniforms of some of the congregation.

Gathered together in the crowd were Nanny and Granddaddy's lifetime friends and family. No one was left uninvited. Even General William H. Simpson, the commanding officer of the U.S. Ninth Army in World War Two was invited. Nanny and Granddaddy knew the importance of the day, not just a wedding, but a turning of the page in their lives. Time to put the war to bed, to look forward. To accept and appreciate those who stood by them, and by whom they had stood, all these years.

From Iowa, Granddaddy's sister Margaret and family were invited and attended. From Georgia, Nanny's mother, Eva "Mommee" (Moore) Palmer, was there. Unfortunately, Nanny's brother Randolph, known to all as the notorious "Uncle Duff," had died the year before, most likely the result of chronic alcoholism.

Despite the many pleas, threats, and schemes, Randolph was whipped by the disease, ravaged by the scourge of addiction. He died a bachelor, divorced from Isabel Amorous Palmer, with whom he had two sons, Martin "Buck" Palmer and George Currell Palmer II. According to my mother, when Randolph and Isabel divorced, Eva "Mommee" Palmer cut all ties with Isabel. Left with no support, in the early 1930s, Isabel moved to Chicago, got a job as a secretary, and married her boss. Her former mother-in-law was furious.

Years later, as a medical student in Charlottesville, Virginia, I came to know "Aunt Isabel" and admired her strength, warmth, and generosity. She welcomed me and the rest of the Van Ness family with open arms.

Among Smiling Jack's family, few stayed away. His Aunt Pauline and her husband, Herschel Schooley, lived in Washington with Aunt Gladys. He was an Interior Department employee and lobbyist. My father's aunts, the sisters of his mother, Ethel, lived in Alexandria, Virginia. Named Aunt Alma, Aunt Clara, and Aunt Ella, they were never ones to miss a good party. Included among the bridesmaids was my father's cousin Betsy Megee. It was Betsy's daughter, the Reverend Katherine Elizabeth "Kitty" Megee Lehman, who was to be the Episcopal officiant and celebrant at the Arlington National Cemetery funerals of both my parents.

As the strains of Handel's "Largo" and Bach's "Air on the G String" played, it was time to begin the wedding. A shared look, a smile between Granddaddy and Tudy, and down the aisle they began a slow walk. At the altar stood Hamilton H. Kellogg, Granddaddy's XVI Corps chaplain. I imagine Chaplain Kellogg's presence was very reassuring; his kind smile and hopeful words had inspired confidence on Christmas Eve just three years before, during the height of the Battle of the Bulge at a service in a bombed-out cathedral in Belgium. That service was a time to reflect, and to resolve to finish the ghastly business of war. In a letter home to Nanny, Granddaddy describes ten soldiers singing "Silent Night" and "Babe of

Bethlehem," writing:

> They were very good. It was evident they had done a lot of practicing for the occasion. I suppose every American's thoughts were far away during the service—I know mine were—but the service and the sermon I know boosted my morale, and I am sure it did the same for the others.
>
> … It was rather impressive to see all the men present, carrying their rifles, carbines, or other arms, and wearing their steel helmets. It is the first time in my life that I have had to wear a pistol when I went to church. But that is what the Germans have done to this world. Hope we can soon destroy these barbarians so that that state of affairs can be ended forever.

Now it was time to put those memories away and celebrate happier times. The march down the aisle, surrounded by his comrades in arms, his family and friends, had to have the feeling of a Roman centurion triumph, absent only a slave whispering in his ear, "All glory is fleeting."

Considering the pain of homecoming with the realization that nothing would ever be the same makes me think that behind Granddaddy's dignified expression lay heartbreak.

Another photo shows that as my mother reached out to Smiling Jack, she gave Granddaddy a quick peck on the cheek before turning away. Jack represented the future in this image, while Granddaddy represented the past. She must have thought, "Daddy will always be there." For the time being, that was true, but it couldn't be true forever.

After the peck on the cheek, Granddaddy stood alone, even though he was surrounded by friends, family, and comrades. Since the end of the war, he seemingly had little purpose in life. A review board occupied some time, but it was a trivial duty compared to taking Winston Churchill over the Rhine. Phone calls were from the bank or the pharmacy, not Supreme Headquarters Allied Expeditionary Force. Now his daughter was marrying. His future would be lonelier, stretching out before him, a time with his wife, his memories, but without his daughter. Their daughter's life with a naval aviator would be a life in a different world than the Army, a world with different dangers and risks, and with airplanes and ships, not horses and artillery. Neither Granddaddy nor Nanny knew that world; they couldn't help the couple except by their presence, just as Chaplain Kellogg's presence on the wedding day must've helped Granddaddy. Chaplain Kellogg's presence was so very reassuring; his kind smile and hopeful words brought a sense of calm to the wedding of my parents.

The election of Senator John F. Kennedy to the Presidency in 1960 was another fairy tale and was seen by many as the passing of the torch from the generation that won World War Two to the next generation. Kennedy's election was welcomed by both the Anderson and Van Ness households. In fact, Nanny had contributed sufficient monies to his election campaign to be invited to the Inaugural Ball in January 1961, hosted by Frank Sinatra. Nanny wanted to meet Jackie Kennedy, an icon of beauty and fashion. A nor'easter storm with plunging temperatures and significant snowfall prevented Nanny and Granddaddy from attending, much to Nanny's disappointment. Regardless, the words of President Kennedy's inaugural address rang true:

> Let the word go forth from this time and place to friend and foe alike, that the torch has been passed to a new generation of Americans – born in this century, tempered by war, disciplined by a hard and bitter peace, proud of our ancient heritage—and unwilling to witness or permit the slow undoing of those human rights to which this nation has always been committed today at home and around the world. And so, my fellow Americans: ask not what your country can do for you—ask what you can do for your country.

We were on the cusp of a "New Frontier" paid for by the sacrifices of blood and treasure during World War Two.

Our delusional world, Washington in the 1960s, blew up in a series of events. The Cuban Missile Crisis of 1962 shook the world, especially the world of military families like mine living in the Washington, D.C. suburb of Bethesda, Maryland. I remember gathering in the downstairs recreation room around the black-and-white television, adjusting the rabbit ears of the cabinet-top antennae to pick up the best signal of the CBS affiliate, Channel 9. President Kennedy outlined the events leading up to the ultimatum to the Russians to remove the missiles pointed at the United States, pointed at Washington.

The implication of his words was too much for my mother, who retired upstairs for a cigarette. My father listened stoically, trying to make sense of the crisis, calculating the odds of open conflict. He would soon be on alert status like all naval aviators, even though it had been decades since he had flown off an aircraft carrier. My mother was upstairs weeping. At nine years old, I had just begun to wrestle with the idea of mortality; now the idea of a nuclear missile detonating over the White House brought a new reality to my world.

Every day at 6 p.m., my father tuned into the news for the latest update on the crisis. What had started with huge fanfare seemed to peter out over the late autumn. The naval blockade organized by SACLANT Admiral Robert L. Dennison, the father of a Landon School classmate of my brother Scott, seemed the decisive act that defused much of the tension and kept my father from needing to leave home with his parachute bag of flight gear situated by the front door, ready for immediate departure.

The assassination of President Kennedy on November 22, 1963 shook Washington and our family to the core. It is a day and time I will always remember well. My fifth-grade class was at recess, outside after our 13:15 lunch, the best part of the day. On the lower-school field at Landon, we were running around, as boys with too much energy always do. We were bumping into one another, wrestling like rambunctious puppies, and yelling and screaming with delight. Then the news began to circulate of Kennedy's assassination. Going from boy to boy, from small group to small group, the horror of it stopped our antics. We didn't know what to do.

I remember seeing our schoolmate Robert "Bobbie" Shriver, the President's nephew, off by himself. He obviously had gotten the word. He was avoiding everyone. Many of us stared, not knowing what to say or do. He avoided eye contact with everyone, instead staring away into space. I remember thinking how sad he looked, and that maybe I ought to say something to him. We played on the same 85-pound football team. But he was a year behind me in school, a deep chasm for a fifth grader to cross, especially at a school like Landon where certain lines were inviolate.

President Kennedy's body was taken to Bethesda Naval Hospital, a familiar place for my family and me. Like so many military families in the area, we used Bethesda for our health care, all free on a space-available basis. After the pathologists completed their work, a mortician from Gawler's Funeral Home embalmed Kennedy's body. Two Marine guards stood watch outside the basement pathology lab as the family gathered in Tower 16 of the iconic Bethesda Naval Hospital. Chief Resident of Internal Medicine Dr. Donald Castell attended the mourners and kept Major General Philip C. Wehle, commander of the Military District of Washington, apprised of the wishes of the next of kin. Kennedy's body was transferred to a mahogany casket for transport to the White House early on the morning of November 23, where it was placed on the replica of the Lincoln catafalque in the East Room. The next day, Kennedy's flag-draped casket moved to the U.S. Capitol Rotunda for a 24-hour period of lying in state, whence he was carried to Saint Matthew's Cathedral and ultimately to Arlington National Cemetery.

Admiral Robert Dennison told the story of attending Presidents Eisenhower and Truman at Blair House the morning of President Kennedy's funeral. While President Truman and his daughter

Margaret waited at Blair House for a ride to Saint Matthew's Cathedral, President Eisenhower called and asked President Truman whether they would like to join President and Mrs. Eisenhower at the service. President Truman agreed and President Eisenhower and Mamie picked them up. After the service, the four of them returned to Blair House. President Truman asked them to come in for coffee. Here is Admiral Dennison's verbatim account, available online at the Harry S. Truman Library and Museum, www.trumanlibrary.gov:

> They accepted, and they all came into the living room of Blair House and sat down.
>
> There were again only a couple of us there. I don't know where the magic came from, maybe President Truman inviting him to come in or maybe because of Eisenhower's thoughtfulness in calling up in the first place, but at any rate they sat down by themselves on a couch and started talking and reminiscing.
>
> They were going along just great and a Secret Service man came in and said, "There's an Army officer at the door who would like to make a statement." I went out to see what this was all about. Well, some officer, it was a colonel, I think, seemed to be embarrassed and didn't know what to do. He said that Mrs. Kennedy sent me over here to make a statement to President Truman, and she understands President Eisenhower is here. She's upset and embarrassed because she forgot to invite these two gentlemen to come over to the White House.
>
> General Eisenhower spoke up first and said, "Please tell Mrs. Kennedy that I understand completely. My wife and I understand it, and it was very kind of her to think of us, but we must get back to Gettysburg, so please present our apology." And Truman spoke up and said, "I feel very much the same way. I appreciate her thoughtfulness and I understand why we weren't thought of in the first place. She has so much on her mind. But I, too, am tired and I've got to rest and I'm sure she'll understand."
>
> I think the colonel thought they would be mad and take it out on him. Instead they just kept on having another

drink and talking, I thought it would never end, but it was really heartwarming because they completely buried the hatchet and you'd think there had never been any differences between them and they were right back where they came in when Eisenhower came back from Europe. So that was the end of their feud and unnecessary hard feelings. And about time, too.

Admiral Dennison's story neglects to show the anguish he must have felt that day. Less than six months before, President Kennedy had decorated him with the Distinguished Service Medal. During the Rose Garden ceremony, Kennedy praised Dennison's leadership during the Cuban Missile Crisis, the closest the United States and the Soviet Union have ever come to nuclear war. In the background of a picture is Admiral Dennison's son, Landon School student Bob Dennison, standing proudly.

My parents took my brothers and me to witness the Kennedy funeral procession. We stood in the crowd at the corner of 17th and Connecticut avenues, the muffled drums and rhythmic footsteps of the military units passing in quiet dignity. As the caisson bearing the flag-draped coffin passed, drawn by matched grays, all one could hear was the snapping of the wind-whipped presidential flag, and the clip-clop of the caparisoned horse, a black stallion named Black Jack. The riderless horse with boots turned backwards was led by PFC Arthur A. Carlson of the 3rd Infantry, the "Old Guard," a simple yet deep tribute to the memory of the heroic "rider."

Nanny and Granddaddy did not come to the funeral procession with us. Nanny was too fearful and overwrought with anxiety; any talk of death and funerals was quickly stifled, as though one could delay the inevitable by avoiding the subject. I never knew my grandparents to attend any Arlington funerals of their friends.

Something else terrible happened about this time. My father had been transferred in the autumn of 1964 to Los Angeles to work on a secret weapons-in-space program. At first, he would spend the week there, returning on the weekend. Red-eye flights got old fast, so he decided to stay in California for the duration of the assignment while the rest of us stayed in Washington. No big deal, right?

My father's separation was depicted as a temporary disruption, but there was more to it than we boys were told. When I overheard arguments and tearful accusations with my mother sobbing and saying, "All I want is to be happy," my stomach would ache as I lay in bed trying to sleep. Whether it started before my father's transfer or after, he was having an affair in Los Angeles.

The day he departed to California, with a jury-rigged cartop carrier for his beloved sailfish boat on his '56 Ford Thunderbird, his manner was jaunty. Not so my mother. She was visibly angry and upset. I was afraid. We stood side by side in the driveway watching my father drive away. No words were spoken.

From an early age, I learned not to express my fears; not to her, not to anybody. Crying just added to her anxieties and was considered a sign of weakness. Likewise, I think she thought she had to be a pillar of strength for her boys, doing for others instead of herself.

So the day my father departed, I threw myself into my chores. Working was something I could do to help out, to create some semblance of normalcy. I took out the trash, raked the gravel driveway, and cleaned out the garage, sweeping dirt and dust in a fury until my allergies swelled my head beyond description.

The following summer, 1965, my mother insisted, despite her fear of flying and my father's dread of the expense of airline tickets and an apartment rental, that we reunite with him on the West Coast for the summer. We stayed in the Quo Vadis apartments in Torrance, California, at the foot of Palos Verdes. My father bought an old Dodge for $50, and Scott would drive us to the beach with the Rolling Stones blaring "Satisfaction" on the radio.

In July, my mother and brothers and I returned to Washington. By this time, my father was having second thoughts and ended the affair. But when he returned to Washington, no one was very happy. The damage had been done.

Unknown to me at the time, Granddaddy laid the foundation for the next chapter, wittingly or not. In the spring of 1967, Granddaddy sat for an interview with us three boys. Scott was a college freshman and John had a new reel-to-reel tape recorder. Scott was to ask questions, John was the technical support, and I simply tagged along. Granddaddy patiently answered Scott's open-ended questions. Nanny interjected from time to time. After thirty minutes, Granddaddy's energy began to flag. He was 77 years old, worn out by hard service, smoking, and whiskey.

The following narrative is derived from that interview, which proved to be one of those "last chance" moments. At the time, we boys thought of the whole thing as a lark, an opportunity to use John's big reel-to-reel tape recorder for something other than playing Beatles songs. We had no idea how precious it was, how remarkable it would be to hear Granddaddy's voice 50 years later. I am so glad to hear not only his voice on tape, but Nanny's in the background, harping on him to tell details of "the latrines used at Ypres" by the British. I even hear their maid, Virginia, come in the front door. A moment caught in time, one of the last tranquil times of my youth.

Later that spring, Granddaddy suffered a major stroke, leaving his speech weak and garbled, his right leg stiff and nearly useless. He could hardly walk without assistance. Instead of strutting down the aisle of the Washington National Cathedral in his pinks and greens with his beautiful daughter on his arm, he struggled up the flagstone path from the street to his front door, aided by my mother or me. Then, in the summer of 1967, all hell broke loose. The Summer of Love in San Francisco, race riots in numerous U.S. cities, and the ongoing Vietnam War shook the foundations of establishment Washington. For the remainder of the 1960s, we preppy boys who had thought our futures lay in hard work and fair play were soon debating the wisdom of sex, drugs, and rock 'n' roll. Like adolescent boys around the country, the boys at Landon began to doubt the wisdom of deferred gratification and self-discipline. So, too, did many of our parents and their friends.

Rejecting conventional wisdom, or so I thought, I resolved to make my own decisions, an example of the youthful egoist whose cruelty and unthinking certainty were in full flood. I never considered going to the Naval Academy. I thought that any of my friends who were thinking about going there or to West Point must be real losers. I turned my back on the institutions that had given so much to my family, and to whom I owed so much. Paul Landon Banfield, the Landon School for Boys, and the military academies of the United States were beneath me. Like the author Ariana Neumann's account in her book *When Time Stopped: A Memoir of My Father's War and What Remains,* I flaunted the rules whenever I could. And I recognized the "unspoken prohibition" that was my own father's childhood.

Now I know better. I know I was young, headstrong, and foolish. Now it is time to make amends, to remember the acts of our forefathers, to celebrate their triumphs, and forgive their shortcomings. It is a way to begin to rebuild our faith in the institutions that serve us so well. For me, it begins with the men of Granddaddy's life, like William H. Simpson and Dwight D. Eisenhower. With their lives, they preached a sermon.

PHOTOGRAPHS

John H. Palmer, left, and brother Samuel B. Palmer, whose Civil War service was anything but, in John's words, "a good easy time, lots of glory, and the stars and bars in the ascendancy." Knoxville, Tennessee (1861)

Samuel Palmer's artwork "Italian Holliday," handdrawn as a prisoner of war, Camp Douglas, Illinois (circa 1864). Photo courtesy of the East Tennessee Historical Society

John B. Anderson, my grandfather, and his first wife, Grace Amoleyetto Wingo, El Paso, Texas (1915)

John B. Anderson with his second wife, Sue Moore Palmer Anderson, my grandmother, taking the traditional bridal "caisson ride" of the field artillery, Fort Benning, Georgia (1924)

Sue Moore Palmer Anderson, member of the Daughters of the American Revolution and United Daughters of the Confederacy, Columbus, Georgia (1924)

Judge George Currell Palmer, who took a chance on young attorney Aaron Cohn, Columbus, Georgia (circa 1925)

Mrs. Isabel Amorous Palmer, sister-in-law of Sue Moore Palmer Anderson. In 1975 in Charlottesville, Virginia, Isabel opened her home and heart to me. Picture from Columbus, Georgia (circa 1930)

Harper Elliott Van Ness, Jr., my father, and his mother, Ethel Stover Van Ness, just weeks before she died unexpectedly, Mexico, Missouri (1919)

Ensign Humphrey Henry Cordes, USNR,
Evanston, Illinois (circa 1942). Photo courtesy
of Jean Cordes Van Ness

Lieutenant David J. Dunigan, USAAF, shown in flight school, served as co-pilot and pilot of B-24s with the 467th Bomb Group, Rackheath, England (April 1944). Photo courtesy of Bruce F. Dunigan

Lieutenant Colonel Malin Craig, Jr., Executive Officer, 106th Infantry Division Artillery, St. Vith, Belgium (December 1944). Photo courtesy of Peter C. Craig

Lieutenant Charles Ashmead Fuller, Jr. of the 69th Infantry Division, sent as a replacement infantryman to the 83rd Infantry Division in the Battle of the Bulge, Ardennes Forest, Belgium (1944). Photo courtesy of Thomas A. Fuller

Charles Carlton Elsbree of Bradford County, Pennsylvania (1944). Photo courtesy of Sandra Lee Elsbree Soni Van Ness

Lieutenant Richard Symmes Thomas Marsh, VMF 321 "Hell's Angels," Solomon Islands (1945). Photo courtesy of Jesse Burgess Thomas Marsh

The Rhine River crossing was the culmination of Operation Plunder. Left to right, facing camera: Major General Anderson, Lieutenant General Simpson, Field Marshal Montgomery (in black beret). Prime Minister Churchill (center), and Field Marshal Lord Alanbrooke (farthest right), Wesel, Germany (March 25, 1945). Photo courtesy of U.S. Signal Corps

*Wasyl Soduk before
World War Two, before
the German occupation of
Ukraine. Monasteretch,
Ukraine (circa 1938).*
Photo courtesy of
Dr. Walter Soduk

Wasyl Soduk family, Monasteretch, Ukraine (circa 1942-43)
Back row, l–r: *Wasyl's father, Konstantin; older brother, Mykola.*
Middle row, l–r: *sister, Marina; mother, Rose; and sister, Nadia.*
Front row, l–r: *sister, Hanusia; sister, Malanka.*
*Wasyl was in German labor camp. When the Russians returned,
Mykola was conscripted into the Russian army, died near
Budapest, winter 1945.* Photo courtesy of Dr. Walter Soduk

Major Aaron Cohn served as S3 operations officer of the 3rd Cavalry Group, Patton's Third Army, Columbus, Georgia (1945). Painting by Ebensee concentration camp survivor, an unknown Polish artist. Photo courtesy of Gail Cohn

Judge Aaron Cohn welcomed us into his home. Left to right: *me, Judge Cohn, and my brother Elliott Scott Van Ness, Columbus, Georgia (2010)*

Warrant Officer John May, Royal Air Force, Lords Cricket Ground, St. John's Wood, England (1940). Photo courtesy of John Tremain May

Newlyweds Sue Moore Anderson Van Ness and Lieutenant Harper E. Van Ness, Jr., my parents, cutting the cake with my father's Naval Academy sword, Shoreham Hotel, Washington, D.C. (December 20, 1947)

Three generations at my mother's wedding reception. Left to right, Sue Moore Palmer Anderson "Nanny," Sue Moore Anderson Van Ness, and Eva Moore Palmer "Mommee," Washington, D.C. (December 20, 1947)

My brother John Anderson Van Ness, left, and me, at Landon School Summer Day Camp, Bethesda, Maryland (August 1959)

At age 12, I surfed at Redondo Beach, California and Indian River Islet, Delaware (summer 1965).

President John F. Kennedy presents the Distinguished Service Medal to Admiral Robert Lee Dennison, Commander-in-Chief of the U.S. Atlantic Fleet and Supreme Allied Commander, Atlantic (SACLANT), Washington, D.C. (29 April 1963). Photo by Abbie Rowe, JFKWHP-AR7857-B, unrestricted

*Colonel Armand Hopkins,
United States Military
Academy Class of 1925,
teaching French at the
Landon School for Boys,
Bethesda, Maryland
(October 1968).* Photo
courtesy of the Landon
School yearbook *Brown and
White*

*Headmaster Paul Landon
Banfield and Mrs. Mary
Lee Banfield co-founded
the Landon School in 1929,
Bethesda, Maryland (1970).*
Photo courtesy of the Landon
School yearbook

Admiral Quintous Crews, Commanding Officer of the National Naval Medical Center, left, decorates me, Lieutenant Michael Moore Van Ness, as my parents, Mrs. Sue Anderson Van Ness and Captain Harper Elliott Van Ness (USN retired) look on, Bethesda Naval Hospital, Bethesda, Maryland (1980)

USS Lansdale *(DD-426) Executive Officer LCDR Robert Morgenthau,* center, *hosts me,* left, *and my brother Scott,* right, *pictured here in front of wall of photos and paintings of Morgenthau's naval service, Manhattan, New York (2010)*

Rear Admiral Paul A. Holmberg was awarded the Navy Cross twice: for "utter disregard for extreme danger" at the Battle of Midway (June 1942) and for "extraordinary heroism and conspicuous courage" in the Solomon Islands (August 1942). Photo courtesy of the Chariton County Historical Society

Honor guard carry flag-draped casket of my father, Captain Harper Elliott Van Ness, United States Naval Academy, Class of 1943, Arlington National Cemetery, Arlington, Virginia (2007)

Display dedicated to my grandfather, Major General John B. Anderson, the Military Aircraft Preservation Society Air Museum, Green, Ohio (2019)

10

Men of the Battle of the Bulge

After the fiasco of Operation Market Garden, an ill-conceived and executed plan of Field Marshal Montgomery to end the war by Christmas 1944, the Germans rebounded. Crouched behind the rivers of Belgium and the Netherlands, the Germans planned a lightning strike to split the English and American armies in the west. Hitler then hoped to negotiate a pact with the Western Allies to combine forces against the Russians and the Bolshevik hordes in the east.

Hitler's plan codenamed "Watch on the Rhine" was conceived in great secrecy. Keeping the Allies in the dark was the key to success and was maintained superbly against all odds. Unlike the campaigns in Normandy and France, where the Allies were greeted with open arms and much useful intelligence by friendly locals waving American and British flags, the Allies approaching the German border in Belgium and Luxembourg were met with stony expressions by sulky villagers.

German morale was rebounding. And while the business of war ground on for the Allied generals, they read the newspapers declaring in bold headlines, "The war will be over by Christmas," (or New

Year's, at the latest). Along the rank-and-file soldiers, they knew the Germans were out on their feet, needing only one last push to go down for the count. At this eleventh hour, on the eve of victory, no one wanted to be the last casualty of war.

On the other hand, everyone wanted to be in on the end, especially the generals. They had waited all their lives for this moment. They had trained regiments, divisions, and armies for this purpose. It was the culmination of a life's work. As quickly as possible, Eisenhower pushed men forward, filling in gaps created by the ever-lengthening lines, gaps the English lacked the manpower to fill. General William Simpson and his U.S. Ninth Army were packed in amongst the fray.

William Hood Simpson

William Hood Simpson was born in Weatherspoon, Texas, in 1888 and graduated West Point in 1909. Known as "Big Simp," he stood over six feet tall, served in the Philippines, and chased Pancho Villa during the 1916 Mexican Punitive Expedition. Simpson was with the 6th Infantry along with his West Point classmate, Captain George Patton. The blazing sun, dust, poisonous water, and rough terrain added significant difficulty to the capture of the elusive Pancho Villa. In fact, for over two years, Pancho Villa evaded capture.

In World War One, Simpson joined the 33rd Infantry Division for the St. Mihiel and Meuse–Argonne campaigns, two parts of the horrific Hundred Days Offensive in the summer and autumn of 1918, campaigns that bled the Germans white, ending the War to End all Wars.

Granddaddy and Simpson were part of a small fraternity, mainly graduates of West Point, the Command and General Staff College at Fort Leavenworth, and the Army War College in Washington. The highest ranks of the Army during World War Two had prepared themselves for nearly two decades. They were the Regulars, who endured low pay and slow promotions, never sure that they would be called on. Pearl Harbor changed all that; now it was time to see if their preparation was sufficient to the task.

In May 1944, immediately before the D-Day invasion, Eisenhower summoned Simpson to his London headquarters. At the time, Simpson had only just arrived in England. Although Simpson had command of the U.S. Eighth Army, there were no combat corps or division commanders assigned to Simpson's command. In fact, U.S. Eighth Army was little more than an organization on paper. This meeting would determine who would be under Simpson's command and what that command would be expected to do.

Eisenhower's first order of business was to change the name of the U.S. Eighth Army to U.S. Ninth Army—in order to avoid any confusion with the famous British Eighth Army. Then, Eisenhower got to the meat of the matter.

Beginning carefully, Eisenhower inquired, "How many corps commanders remain in the United States?"

Simpson's reply, "About nine."

"Who are they?" Eisenhower asked.

As Simpson named each one, Eisenhower wrote down the names. Eisenhower had a longstanding agreement with Army Chief of Staff Marshall that Eisenhower would select his generals, with little discussion or fanfare. Eisenhower had full control of the process and exercised that authority ruthlessly. Ever confident of his own judgment, Eisenhower crossed out five of them. Down to four names, he gave Simpson the last word.

"I can accept any of these four; as Army commander, you can name three out of these four." Granddaddy was one of the three corps commanders named by Simpson.

Notified of the selection, Granddaddy and his XVI Corps expected to join the fight in the summer of 1944. However, several events delayed the arrival of the XVI Corps. The first was the terrible summer gale that destroyed the artificial harbor, the British

Mulberry system, off the Normandy coast near Caen. The destruction of the artificial harbor meant that the buildup of British forces slowed. The Americans would not be able to count on the Mulberry to speed arrival of new units. Instead, the Americans needed to capture either Cherbourg or Brest or both to support their operations in Normandy, and that would take time.

After a three-week siege, the Germans surrendered the port of Cherbourg to Major General "Lightning Joe" Collins' VII Corps on June 26, 1944. The harbor facilities were destroyed by the retreating Germans, the channel into the harbor clogged with multiple sunken hulks, barges, and ships. Clearing the harbor and restoring it to use required weeks of intense effort. Even when Cherbourg was back on line, German U-boats prowled the port entrance, reducing the efficiency of Allied supply efforts.

Brest, on the Brittany peninsula, proved another hard nut to crack. Major General Troy Middleton's VIII Corps of Patton's Third was given the assignment. As Patton sped east into France, Middleton found himself something of an orphan, forgotten by his commander intent on the capture of Paris. It was not surprising. Patton and Middleton had a strained relationship going back to events in Sicily. At the time, William Mauldin was a correspondent in Middleton's 45th Division. Mauldin created the cartoon characters Willie and Joe, dog-faced soldiers who endured the rigors and danger of combat, all the while making Patton the butt of their jokes. Patton was not amused by Mauldin's cartoons. When Middleton defended Mauldin, Patton was irritated with his subordinate. A year before, the public outcry to Mauldin's cartoon of Patton kicking Pvt. Charles H. Kuhl in the pants and slapping Pvt. Paul G. Bennett had nearly ended Patton's career. The swastika Mauldin emblazoned on Patton's boot was particularly galling. Infuriated by the slapping-incident cartoon, Patton did not take kindly to anyone defending it and never forgot Middleton's refusal to curb Mauldin's work.

Brest was not going to fall without a real fight. The German commander, Generalleutnant Hermann Ramcke, took to heart his orders to hold at all costs. With Patton preoccupied with Paris, Eisenhower arranged for Simpson and his Ninth Army to assume command of the all-American forces attacking Brest, including Middleton's VIII Corps. For Simpson and Middleton, it was a happy reunion of two war horses who had fought together against Pancho Villa in 1916 and the Germans in the Meuse–Argonne Offensive in 1917. In a bitter fight, Brest held out for weeks, finally surrendering on September 19, 1944, but not before Ramcke's 36,000 German soldiers completely wrecked the port. Simpson ensured that Middleton received credit for the victory, making sure Middleton received Ramcke's surrender in person. In turn, Middleton gave the city back to its mayor, in a formal ceremony acknowledging the end of Nazi occupation.

The fall of Brest meant that Simpson's army would now expand. Granddaddy's XVI Corps moved quickly from Fort Riley, Kansas, to Gourock, Scotland, and over Omaha Beach to Barneville, France, on the west coast of the Cherbourg Peninsula. Assigned a supportive role, for the next two months, Granddaddy's XVI Corps funneled men and material from Cherbourg to Hodges' First Army and Simpson's Ninth Army. Granddaddy visited Brest several weeks after Simpson and Middleton's conquest, and wrote:

> The total destruction of that city, its port facilities and harbor cannot be described. It had to be seen to realize its complete and total waste. The poor people were drifting back to it at the time I visited it, but where they were to live, I don't know—in cellars and caves, I suppose—as they are doing in so many of the destroyed cities all over Europe. How one man could be responsible for so much destruction and suffering, not only among the conquered nations but among his own people, and escape an assassin's bullet is difficult to understand.

Granddaddy also had the opportunity to witness first-hand the ruthless nature of command politics. His friend, Leroy H. Watson,

USMA 1915, was a favorite of Marshall, Eisenhower, and Bradley. He was given the U.S. 3rd Armored Division for the Normandy campaign, under the command of VII Corps commander General J. Lawton Collins. Collins was unhappy with the 3rd Armor's performance in the hedgerows of the Cotentin peninsula. About the time Granddaddy arrived in Barneville, Watson was relieved. Watson was given the chance to either return to the United States to a training command with full rank and privileges, or remain in France as a colonel on Bradley's staff. He chose to stay in France to try to earn his way back. For Granddaddy, his friend's experience was a reminder that there was always someone watching and always someone waiting in the wings.

For several weeks, Anderson's XVI Corps received, equipped, and transported army units of various kinds to the front, a thankless service job they were glad to give up in late November. Moving into Belgium, the number of men in Anderson's XVI Corps swelled in preparation for combat. On the morning of December 16, the 75th and 79th Infantry Divisions were already under his command. The 106th Infantry Division was scheduled to come over at noon. These three divisions would go into the front lines to relieve the British XII Corps.

Events transpired to foil those plans. By the afternoon of December 16, both the 75th and the 79th "were gone," to quote from Granddaddy's letters. The men of the 106th remained in General Leonard T. Gerow's V Corps of General Courtney Hodges' U.S. First Army. Granddaddy had to wait for the crisis to pass. In the meantime, battle raged.

Lieutenant Colonel Malin Craig, Jr., Executive Officer, 106th Infantry Division Artillery

Lt. Colonel Malin Craig found himself in the middle of the fight. As the Executive Officer of the 106th Infantry Division's Artillery, his job was to inspire and lead other men in the risky business of combat. His immediate senior officer was Brigadier General Leo T. McMahon. The two of them were responsible for the 589th, 590th, 591st, and 592nd Field Artillery Battalions, each with a headquarters company. In total with the four field artillery battalions, Craig was responsible for 36 truck-drawn 105mm cannon, 12 tractor-drawn 155mm cannon, and over 2,000 men armed with various weapons like bazookas (forty per battalion), .50 caliber heavy machine guns, and M1 Garand battle rifles and carbines.

Craig was a graduate of the United States Naval Academy, Class of 1924. At graduation, he made the unusual but not unprecedented switch from the Navy to the Army. Coming from a family with a proud tradition of service, the switch was understandable. Colonel Craig's father was General Malin Craig, whose distinguished career included World War One service in the murderous 1918 Meuse–Argonne campaign. Now his son, Malin Craig, Jr., was in a desperate fight. True to his family's traditions, the younger Craig rose to the occasion. Colonel Craig's sons Peter and Joe were Landon School boys, part of that loose-knit D.C. community of veterans whose stories I am just now discovering.

Within 48 hours of the start of the German attack on December 16, 1944, the two northernmost regiments of the 106th, the 422nd and the 423rd, were surrounded and badly cut up. Effective communication with these battalions was lost, so Craig set out from St. Vith to find them. Craig moved the division's artillery command post from St. Vith west to Poteau, about 7 miles up a twisty, muddy lane. From Poteau, the artillery would be in a better position to direct fire on the Germans attacking St. Vith.

Craig led a column of 25 vehicles from the headquarters' group into Poteau, setting up an artillery command center. While the headquarters group assembled, telephone communications with General McMahon in St. Vith were lost, likely cut by German artillery fire, hampering the withdrawal of the actual artillery batteries. Radio communications were unreliable, overwhelmed by countless operators and vulnerable to static, misunderstanding, and interception by the enemy. The Germans were rendering many radio frequencies useless by jamming. Messages sent "in the clear" were avoided, as they simply alerted the Germans of American intentions. Secure, coded radio message traffic was possible, albeit slow. Civilian telephone lines could be used, but one never knew if the Germans were listening.

Under the circumstances, communication was clearest and most reliable face-to-face. As darkness settled over the Ardennes on December 18, the third day of the battle, Craig headed back into St. Vith. He needed to confer with General McMahon. He also needed to round up any remaining elements of the 106th's artillery near St. Vith. On the Poteau–St. Vith road, at about 0200 on December 19, Craig found the remnants of the 589th Artillery Battalion commanded by Major Arthur C. Parker III. Craig also found elements of the artillery headquarters' group and its most senior officer, Lt. Col. Burtis L. Fayram. Craig ordered Fayram to move everyone west toward Poteau, into a useful firing position. He wanted the guns to face northeast and the vehicles off the road, but facing it, ready to go either direction, east or west. Fayram appeared to understand.

Craig then headed out into the darkness to the east, toward the sound of the guns in St. Vith. Craig reported to General McMahon, who re-affirmed the need to get the remainder of the division's artillery out of St. Vith and into a firing position west of the St. Vith, preferably in the crossroads at Poteau. Having taken care of the 589th, Craig turned to the 592nd. Radio and telephone communication was impossible, so Craig headed out to the 592nd's last known position in the village of Schlommefurt. As the son of a former Chief of Staff of the Army, Craig knew the drill, "March to the sound of the guns." He likely would also have agreed with Henry Foley, who said years later at a GCSE National Curriculum conference:

No matter how much you love your country, it won't love you back. No matter how much you believe in the military, it doesn't believe in you. Your duty is required, and your loyalty is expected. If you scorn these traditions, you'll be branded as dishonorable. It is a difficult and thankless, and sometimes perilous, commitment fitted only for a few hard and dangerous men. Any who answer the urgent call of the drums, and march toward the sound of distant guns, must be willing to die unsung, unwept, and unknown.

Sniper fire was intense out on the road. Craig pressed forward and found the 592nd in the woods just east of Schlommefurt. They were preparing to destroy a disabled tractor, the destruction of which would have blocked the very road they needed to use. Craig got the men of the 592nd to follow him and head northwest, with the intention of joining the headquarters group and the 589th on the road east of Poteau, the site of the new firing position.

Moments before Craig's arrival, a motorcyclist, most likely a German soldier in an American uniform, one of Otto Skorzeny's "Trojan Horsemen" of Operation Greif, had raced through the site, dropping a flare along the road. Within seconds, deadly German artillery and mortar fire began to fall, killing several and scattering the rest. Craig with the 592nd in tow arrived to find the planned site a mass of confusion.

After the barrage, the Germans attacked. Small-arms fire and machine guns scattered the headquarters group and the 589th. Unit cohesion and discipline evaporated. Craig stepped in and restored some order. He put everyone who could drive behind the wheel of one vehicle or another, backing and filling. Adding to Craig's difficulties was Fayram's failure to deploy the vehicles facing the road. Fayram and elements of the headquarters group became separated and joined men of the 771st Field Artillery Battalion, an VIII Corps artillery unit, in a hasty retreat westward.

Craig led the ragtag force south toward Bovigny, small-arms fire from both flanks harassing the convoy. Surviving this gauntlet run,

Craig positioned the three artillery pieces of the 589th and the entire complement of cannon of the 592nd into a satisfactory firing position along the Vielsalm-Bovigny road. Craig then led the support vehicles of the two battalions into Vielsalm. Craig's efforts to consolidate the forces of the 106th artillery command were soon complicated further.

While Craig was heading toward Vielsalm, Colonel Herbert W. Kruger, the Commanding Officer of the 174th Field Artillery Group of the VIII Corps Artillery, came upon the 589th. He informed Major Arthur C. Parker III that Parker and the 589th were under his command, a voice order of the Commanding General of the VIII Corps, "VOCG VIII Corps." Kruger wanted the 589th to protect his own 174th Field Artillery Group's five battalions from German tanks reported to be gathering nearby at Cherain.

At dawn on December 19, Kruger dispatched the 589th to Salmchateau, and then further west to Baraque de Fraiture, again to protect the 174th from German tank attack from Samree. No tanks appeared.

Parker's men were running low on gas, food, and ammunition. Despite being part of Kruger's command, Kruger and the VIII Corps were providing little support for the 589th—no food, water, or ammunition. When Parker sent trucks into Vielsalm, to the 106th looking for help, Craig learned the 589th had been commandeered by Kruger. It did not take long for the 589th to return to the 106th.

Even as the 589th returned to the fold, Craig realized the 592nd and headquarters group were nowhere to be found. We know the 592nd had set up firing positions along the Vielsalm-Bovigny road, near the 589th, south of Vielsalm. Here's what happened next. After the 592nd was established in position, Lt. Col. Richard E. Weber, the acting commander of the 592nd, sent CWO James B. Bennett of the 592nd west to Marche to search for supplies of gasoline. There Bennett ran into Lt. Col. Fayram, detached from the unit since the

German attack. Resuming a command role, Fayram ordered Bennett to return to the 592nd and decamp "to the west, toward La Roche and Marche." According to Charles MacDonald writing in *A Time for Trumpets*, Bennett passed on this order to Weber, who interpreted the order as legitimate; after all, Fayram was the S-3 (Operations Officer) of the 106th Field Artillery Headquarters Staff. Communications being unreliable, Weber had no way to confirm Fayram's order. So off went the 592nd, first through La Roche, then La Marche-en-Flamenne, heading out of the fight, heading west toward the Meuse River. According to Army historian Colonel R. Ernest Dupuy, the movement of the 592nd deprived the 106th of its medium artillery for three critical days.

Informed of the 592nd's departure, Craig took it upon himself to find the errant unit. Chasing the 592nd west, through La Roche and into the village of La Marche-en-Flamenne, 25 miles west of Bovigny and Vielsalm, Craig came upon a Lt. Col. Harvey R. Fraser of the 51st Combat Engineers. Fraser reported to Craig the 592nd had come through, on its way to Dinant, another 30 miles to the west, along with elements of the headquarters group commanded by Fayram.

Craig soon caught up with Fayram and the headquarters group in two trucks commandeered from the 423rd Infantry. Turning them around and driving all night, Craig escorted them back to La Marche-en-Flamenne, before heading out again, this time to find the bulk of the 592nd. After another harrowing trek on snow-covered, muddy, and jammed roads, Craig found the 592nd 15 miles west of Dinant, in Rosee, west of the Meuse River. A few choice words, and the 592nd was headed back to Erizee, and then to positions near Commanster, a half-mile from Vielsalm, moving faster than Craig had ever seen an artillery outfit move, to reinforce the 591st. Lt. Col. Fayram's decisions and orders were subsequently investigated; he was relieved of command.

The artillery of the 106th was now back in the fight, contributing mightily to the turning of the tide at Parker's Crossroads and Manhay. Colonel Craig, for his achievements in the face of overwhelming odds and for showing personal initiative in the re-establishing the integrity of the 106th's artillery, was awarded both the Bronze Star Medal with "V" and the Legion of Merit.

Berga

Many of the 106th soldiers captured in the initial assault were interned, not in military prisoner of war camps, but in Nazi slave labor camps like Berga. The 2003 movie *Berga: Soldiers of Another War* dramatizes the maelstrom in which 350 American soldiers were caught. Among them was author Kurt Vonnegut, who would write about his experiences in *Slaughterhouse Five*. Slave labor camps were not governed by the rules of the Geneva Convention. American soldiers, mixed among Russian, Polish, and Ukrainian laborers, received no Red Cross packages of food and clothing and were afforded no mail privileges. Starvation rations kept the men alive to work in caves and other underground facilities dug for protection against Allied aircraft, and to toil in underground caverns designed to protect a planned jet-fuel factory.

As the Allies drove deeper and deeper into Germany, conditions at Berga deteriorated. Survival depended on vigilance, ruthlessness, and the buddy system. Starvation drove men to steal food and clothing from one another in a desperate effort to survive. While some slept, others remained awake and "on guard" to protect the meager supplies of food that often meant the difference between life and death. Later, the German guards abandoned Berga and marched the prisoners in ragged columns deeper into Germany to escape the advancing Allies. All semblance of order broke down as guards shot stragglers and prisoners realized they were at the end of the war and at the mercy of crazed guards. Finally, the guards simply disappeared, leaving the prisoners to their own devices.

American soldiers interned at Berga lost 40–45% of their body weight, many starting out at 175 pounds and coming back into American lines at 100 pounds. With proper nutrition and medical care, they regained their physical health, but the psychological wounds were deep and endured.

Years later, Admiral James Stockdale wrote about his ordeal as a prisoner of war in Vietnam. For six and a half years, he was detained in the Hoa Lo prison, infamously known as the "Hanoi Hilton." When the North Vietnamese selected Stockdale as a propaganda tool to parade in front of the world as an example of their humane treatment of prisoners, he cut his own scalp and beat his face to a pulp. Later, he slit his own wrists to show his captors he would rather die than capitulate. When Stockdale was asked how some survived and others did not, he replied that faith "in a just end" was absolutely necessary. Those who died were too often hopeless optimists whose dreams of "home by Christmas" were too often dashed by reality.

I have had the privilege in my medical career to know some of these men. By offering little more than a sympathetic ear, they seemed reassured by my personal and family's history of military service. They told wretched stories of their ordeal—first in combat, and then in captivity—sharing stories of petty theft, overwhelming fear, and near starvation. The degradations of dysentery, filth, and disease compounded the misery of the American soldiers, especially the third of them with dog tags stamped with an "H," meaning "Hebrew." However, regardless of religion, the captives at Berga were considered slaves, valuable only so long as they could work. They would hold the heavy jack hammers that drilled holes to contain dynamite to explode the rock face of the underground tunnels. Once the charges were placed, the men tried to find shelter from the blast. Oftentimes, shelter was only behind a wooden board or around a corner and behind a rock—and never far away.

The prisoners endured the concussions of the blasts, breathed the cordite fumes, and then returned to the rock face to labor in the thick clouds of rock dust created by the confined explosions. Chunks of rock were dislodged, too big to be hoisted into the battered metal hopper cars of the narrow-gauge railroad. The slave laborers wielded sledgehammers to break up the boulders, and then shoveled, or hand-carried the remnants to the rail siding for removal, all the time inhaling the choking rock dust. It is no wonder so many died of respiratory illnesses—asthma, bronchitis, diphtheria, and pneumonia. Typhus, skin ulcers, and meningitis ravaged the men. And when they were simply too exhausted, too sick, or too weak to go on, when their suffering exceeded the limits of human endurance, they were beaten, or executed.

North of the U.S. 106th, German Kampfgruppe leader Joachim Peiper had his lead tanks in open tank territory, heading west into the heart of the U.S. First Army. Needing only to get across either the Salm or Ambleve Rivers, he had reason to believe his 237 armored vehicles and 8,000 panzer grenadiers would reach the Meuse River, splitting the Allied armies in two. His most deadly weapons, 60-ton Royal Tiger and Panther tanks, were thirsty for American gasoline. For the time being, mobile refueling trucks were keeping Peiper on the prowl, but he needed to find the American supply dumps with their vast quantities of gasoline to continue his effort. Moving up and down the American lines, probing among the numerous villages and hamlets with their stone and timber bridges over the creeks and ravines of the Ardennes, time was beginning to run out. He had already penetrated the crust of the American defenses, brushing aside elements of the 99th Infantry Division and the 14th Cavalry Group.

On the morning of December 18, 1944, other than geography, there was little to stop Peiper's column. While the 82nd Airborne Division, under Major General Gavin, was racing to establish a defensive line farther to the west, it was the men of the 1111th

Engineer Combat Group under the command of World War One veteran Colonel H. Wallis Anderson who alone faced the advancing Germans.

On December 18, the 1111th Engineer Group included the 51st Engineers under Lt. Colonel Harvey Fraser and the 291st Engineers under the command of Lt. Colonel David E. Pergrin. Pergrin was responsible for security of the 1111th Engineer Group and had his own headquarters in the village of Haute-Bodeux, just west of Trois Ponts. One company of Pergrin's command, Company A under Captain James H. Gamble, was further west in Werbomont, while Company B was in Malmedy, and Company C in La Gleize. On the morning of December 16, Gamble began to move his men of Company A forward, to support the 9th Armored Division's planned move to the Roer River dams. Gamble's men moved through the village of Ambleve and had sawmills operating in both Montenau and Born, halfway between Malmedy and St. Vith. They were not to stay long.

Unbeknownst to the Engineers, they were in the immediate path of Kampfgruppe Peiper. Cutting down pine trees of the Ardennes forests and planing them into useful lumber for construction of winter barracks for American troops was soon to be forgotten in the desperate effort to stop Joachim Peiper from reaching the Meuse River.

The first inkling that something was afoot occurred before dawn on the morning of December 16. Four large-caliber artillery shells from the big German railroad guns landed in Malmedy. The guns had a 30-mile range and fired 280mm shell. Designed to instill terror, the railway guns were limited to 20 to 50 rounds before the rifling deteriorated and accuracy lost. Difficult to conceal and thus vulnerable to air attack, the clouds and fog of a mid-winter inversion concealed the terror weapon during the attack. The shell bursts were terrifying. More so was the steady stream of American soldiers retreating from the east. No one knew the extent of the German attack, nor the best means to thwart the enemy advance. In his book *A Time for Trumpets,* Charles B. MacDonald notes that in Baugnez, near Malmedy, First Lieutenant Frank W. Rhea, Jr., of Company B witnessed the flow of men and material heading west, remarking later that the town looked "like a giant anthill somebody had stirred with a stick," vehicles pouring in on two sides and out the other two.

When Pergrin heard the reports via the Company B commander Capt. John T. Conlin, Pergrin set out for Malmedy to see for himself. Malmedy was in pandemonium as the evacuation of a field hospital, replacement depot, quartermaster and ordnance units, and civilians clogged the streets. Taking the initiative, Pergrin ordered Company C at La Gleize forward to Malmedy, dropping a squad at both Trois Ponts and Stavelot.

At Stavelot Company C of the 291st Engineer Combat Battalion began the fight that ultimately stopped Pieper's advance. A lucky bazooka shot disabled the lead German tank. Intense small arms and machine gun fire confused the Germans into thinking a larger force held the village. A stealthy withdrawal gave the Americans time to set explosives for destruction of the bridge over the Ambleve River. The exhausted Germans stopped for the night. The Seventh Armored Division passed through Stavelot on their way to St. Vith. The Germans waited.

While Kampfgruppe Peiper re-grouped, it was time to blow up the bridge at Stavelot. Soon after midnight, the engineers hit the plunger. Nothing happened. How was that possible? The charges had been set carefully by experienced engineers, experts with six months' experience in the field from their time in Normandy to the battle in the Ardennes. What they did not know was that two Greif operatives were roaming amongst their ranks.

The next morning, Kampfgruppe Peiper rolled toward Stavelot; the bridge over the Ambleve still intact. Kampfgruppe Peiper passed through Stavelot and headed for Trois Ponts. The engineers had the bridges at Trois Ponts wired with tons of TNT, but no one knew the answer to the question, "Had the Greif operatives disabled the charges there, too?"

While the engineers had mines, primacord, and TNT in great supply, what they lacked were men—men to set charges, check charges, detonate charges—all the while fighting against the determined German advance. At Trois Ponts, Col. Anderson had to stop Kampfgruppe Peiper with a force of 140 men, 8 bazookas, 10 machine guns, and a 57mm anti-tank gun manned by a four-man gun crew from the 526th Armored Infantry.

Anderson and his men watched the German advance through the morning mists. As the first tank emerged, the gun crew opened fire with its 57mm anti-tank gun, knocking out a Panther on the road just east of the Ambleve River bridge (see Map #1). The Germans responded with withering automatic-weapons fire, killing all four men of the gun crew. The Germans pressed forward. They were relentless and crazed with Pervitin, a methamphetamine that produced higher energy levels and euphoria while decreasing the need for food or sleep. Soldiers on both sides of the conflict used amphetamines to increase endurance and stamina. In the Battle of the Bulge, the Germans used Pervitin extensively, even mixing it with chocolate for ease of administration. It was now or never for Kampfgruppe Peiper, the spear tip of the German offensive Wacht am Rhein (Watch on the Rhine), and for Hitler's last roll of the dice to win the war in the face of overwhelming odds.

Retreating west over the curving Ambleve River, Anderson's men dodged the intense small-arms fire. They had to get to the detonator and blow the bridge. Cover fire came from the bazookas and machine guns arrayed on the west bank of the Ambleve. With the anti-tank crew dead, there was little left but to escape and then blow the bridge. The Germans could see the last of the hunched-over soldiers escape over the stone bridge. And then, in the face of the advancing Germans, the bridge was destroyed, nearly vaporized by the double-charges set, checked, and double-checked. The Germans were so close, like the Confederates at Bloody Angle on the third day of the Battle of Gettysburg, so nearly into the rear of their demoralized and confused enemy. Trois Ponts was the high-water mark, the tide to rise no farther.

With the explosion, Colonel Peiper raged, "Those damned engineers!" Peiper knew his objectives, the First Army supply dumps were just beyond his reach. Not easily foiled, Peiper sent three German light half-track reconnaissance vehicles seeking other routes, ones that might not be shown on his maps. Two of them drove down a lane, over a string of eight anti-tank mines laid across the road by Private Johnny Rondenell of the 2nd squad. The anti-tank mines did not explode. Puzzled, Rondenell came out of his hiding spot to reposition the mines, which destroyed the third half-track, flipping it on its back, blocking the road.

The first two German half-tracks continued into the hamlet of Forges and over the Ambleve. As darkness fell, the halftracks continued to reconnoiter, turning south toward Habiemont, to be destroyed by troops of the newly arrived U.S. 30th Infantry Division. Peiper never learned that just around the bend was another bridge suitable for his Panzers.

Peiper headed north, probing toward Habiemont, where there was a bridge over the Lienne Creek. If he could cross that bridge, he'd be in the clear. S/Sgt. Edwin Pigg and Sgt. R.C. Billington, the only line troops left in Company A of the 291st Engineers, had other ideas. They wired the bridge with 2,500 pounds of TNT, including a back-up charge should the first fail. The engineers lay in wait for the Germans. While they waited, the two sergeants were joined by their platoon commander, Lt. Alvin Edelstein and Cpl. Fred W. Chapin, Asst. 3rd squad leader. They even had a visit from Major General James Gavin of the 82nd Airborne, newly arrived from U.S. First Army headquarters to set up the 82nd's command post in Werbomont. Although the timing of the bridge's destruction was important, its complete destruction was more so.

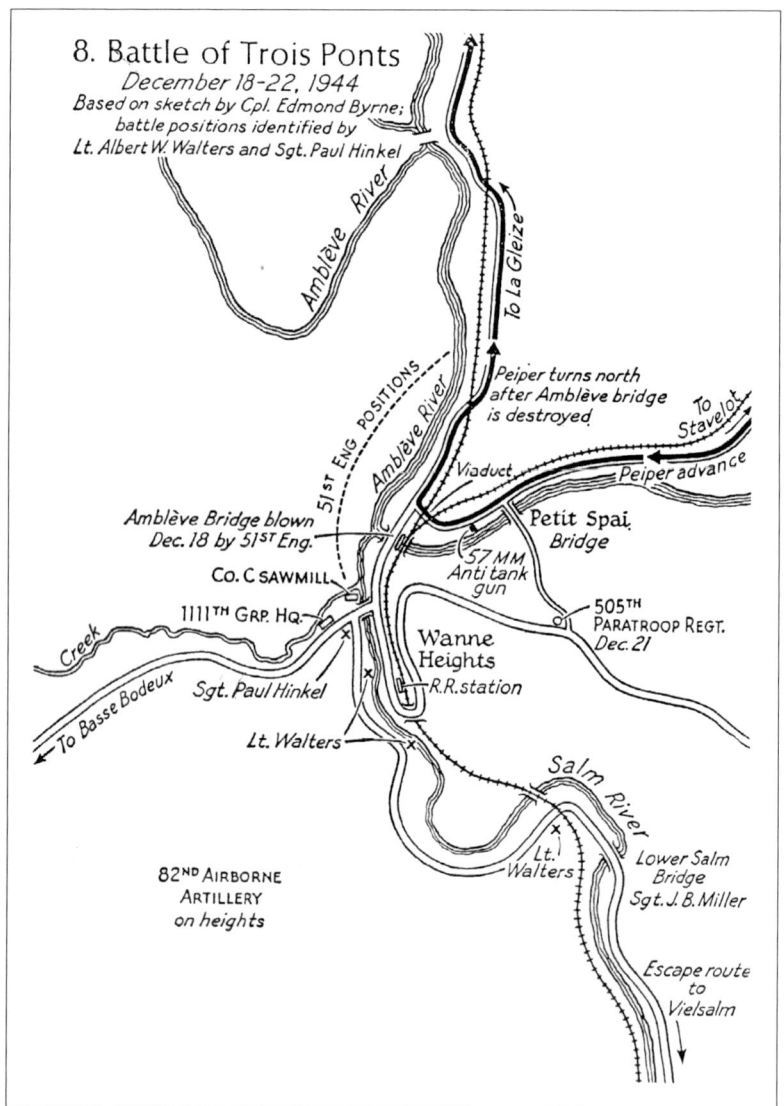

Map #1. Battle of Trois Ponts: The men of the 51st Engineer Battalion of the 1111st Engineer Group were all that remained between German Kampfgrupppe leader Joachim Peiper and the Meuse River, Trois Ponts, Belgium, December (1944). Map courtesy of Schiffer Military/Aviation History, Atglen, PA

It was "High-Noon at the OK Corral." A German Tiger Royal tank, the behemoth sporting an 88mm cannon, led the way. No tank was more feared by the Allies than the Royal Tiger, its thick, sloped front armor nearly impervious to the anti-tank guns and tank cannons of the Allies. The Royal Tiger's 88mm cannon had range and velocity that made it superior to any armored opponent. At the moment the Royal Tiger appeared, all thought of waiting for the Germans to come closer vanished. Chapin looked to Lt. Edelstein, who gave the signal to blow the bridge. In an enormous explosion, the 180-foot span disappeared into the river.

Peiper had shot his bolt. He bivouacked his command for the night at the town of Stoumont, less than 25 miles from his objective, Liege. Had he been able to cross the Salm River at Trois Ponts, the Ambleve north of Trois Ponts, or Lienne Creek at Habiemont, First Army supplies would have been his for the taking. A straight run to Liege, where signal, road, and rail center were piled high with more supplies, and even an advance toward British bases at Brussels and the port of Antwerp, would have had huge strategic consequences. Without the bridges, Kampfgruppe Peiper was trapped. Without the supplies, Kampfgruppe Peiper could not move. Kampfgruppe Peiper was cut up and destroyed over the next two weeks by General Matthew Ridgway's XVIII Airborne Corps and "Lightning Joe" Collins' VII Corps.

Robert Merriam, the historian for the 7th Armored Division, wrote of the heroics of the engineers under Pergrin's and Anderson's command in his 1947 book, *Dark December:*

> Here was a case where the fate of Divisions and Armies rested for a few brief moments on the shoulders of a handful of men: first, at the town of Trois Ponts, and then, only hours later, with another, smaller, handful of men at the bridge just east of Werbomont. Had either of these groups failed in their job (and the temptation to run away must have been very great), the probability that Peiper would have got to the

Meuse River the next morning, behind Skorzeny's "Trojan Horse," would have been very high.

While the 106th Infantry Division was slowing the advance of Sepp Dietrich's Sixth Panzer Army in the north, and the 1111th Engineer Group was stopping Peiper's advance at Trois Ponts and Habiemont, Major General Norman Cota's 28th Infantry Division slowed the advance of von Manteuffel's Fifth Panzer Army. The climax of the battle was the siege of Bastogne.

11

Bastogne

As reports trickled in throughout the first day of the Battle of the Bulge, Eisenhower and Bradley focused on the developments in the north. They thought the Germans were simply trying to disrupt the American offensive against the Roer River dams. Just in case something larger was afoot, Eisenhower ordered Bradley to get help to General Middleton, Commanding General of the VIII Corps of Hodges' First Army. Bradley hesitated; General Patton was preparing a big offensive. Bradley's orders to take an armored division away from Patton were not going to be received well.

Bradley made the call himself. Patton pushed back, belittling Middleton's request for the 10th Armored Division, saying Middleton could handle it himself.

Bradley reiterated his order, "I hate to do it, George, but I've got to have that division. Even if it's only a spoiling attack as you say, Middleton must have help." Patton was irate.

By way of contrast, when Eisenhower ordered Bradley to send Simpson's 7th Armored Division to First Army, Simpson's response was professional. Without objection, Simpson calmly relayed to his XIX Corps commander, Major General Raymond S. McLain, that 7th Armored Division under Brigadier General Robert W. Hasbrouck was to head south to St. Vith, to General Hodges' First Army.

Later that day, Hodges bypassed Bradley and called General Simpson directly. Hodges had good reason; he needed help and needed it quickly. More than the other senior generals, Hodges recognized the primary threat posed by Kampfgruppe Peiper to the 2nd and 99th Infantry Divisions in the Losheim gap, immediately to the north of the 106th. Hodges asked Simpson to add Major General Leland Hobbs' 30th Infantry Division to the 7th Armored Division. Simpson did not hesitate.

With Hobbs' 30th under his command, Hodges knew he could shift other forces to the south, to aid General Middleton, who was being hard pressed by the Fifth Panzer Army. Hodges gave Middleton Major General John W. Leonard's 9th Armored Division. But he knew Middleton needed more. Simpson had nothing else to give. Hodges wanted the two airborne infantry divisions that constituted Eisenhower's strategic reserve. Eisenhower reluctantly agreed to send the two divisions into the fray.

The 82nd and 101st Airborne were coming, part of the XVIII Airborne Corps under Major General Matthew B. Ridgway. However, it would take some time, time Hodges did not have. Middleton needed help *now*. Since Ridgway was in England and Major General Maxwell D. Taylor of the 101st was in Washington, D.C., Major General James M. Gavin of the 82nd would be the acting corps commander, at just 37 years old the youngest American corps commander in the field. Wasting no time, General Gavin set out at midnight on December 17 for General Hodges' First Army Headquarters at Spa, Belgium.

As Gavin raced north with the 82nd, other American forces headed for Bastogne. The Volksgrenadiers and Panzergrenadiers of von Manteuffel's Fifth Army were only 10 miles to the east. Only Major General Norman "Dutch" Cota's 28th Infantry Division stood in the way. They were the men of the "Bucket of Blood" division, so named for its distinct red shoulder patch that resembled the keystone design of the state of Pennsylvania, home to many of its soldiers.

General Cota was a true combat leader, with a distinguished record of heroism. On D-Day, he was the assistant division commander of the 29th Infantry Division, the highest-ranking officer on Omaha Beach and, at age 51, the oldest. As the men at D-Day were pinned down on the beach, so graphically depicted by Tom Hanks in the movie *Saving Private Ryan*, Cota recognized the disaster in the making. Cota rallied the men of the 5th Ranger Battalion to advance through a breach in the seawall with the words, "... if you are Rangers, then get up there and lead the way." There are numerous eye-witness accounts of General Cota's leadership on D-Day. While battlefield noise and confusion make a verbatim record nearly impossible, Stephen Ambrose in his book *D-Day June 6, 1944: The Climactic Battle of World War II* quotes Captain John Raaen, a West Point classmate of Cota's son, as saying, "You men are Rangers and I know you won't let me down." There was another witness at Colonel Schneider's CP, Sergeant Fast, who recalls the words, "I'm expecting the Rangers to lead the way." These accounts and others are from the personal oral histories now housed at the Eisenhower Library in Abilene, Kansas. The Rangers followed Cota over the seawall. Then saying, "Gentlemen, we are being killed on the beaches; let's go inland and be killed," Cota got the men of the 116th Infantry Regiment of the 29th Infantry Division off Omaha's Dog Beach, opening an essential "draw" for other men to follow. For these actions, he was awarded the Distinguished Service Cross.

Cota was promoted to command of the 28th Infantry Division in August 1944. Only hours before, Brigadier General James E. Wharton, a ROTC graduate of the New Mexico State University, had replaced General Lloyd Brown as the commanding officer of the 28th but was killed by a German sniper on an inspection tour of the 112th Infantry Regiment near Sourdeval, Normandy, France.

On December 16, 1944, Cota had only the 28th Division, with three U.S. infantry regiments, the 112th, the 109th, and the 110th, against both the 2nd and 116th Panzer Divisions, supported by the 352nd Volksgrenadiers, the crack 5th Parachute Division, and the 26th Volksgrenadiers. Six thousand or so infantrymen, shocked by the initial German assault, against two fresh, reinforced armored divisions determined to drive the Americans from the borders of their Fatherland. It was not a fair fight, but not in the way one might think.

On the 28th Infantry Division's left flank, the 112th Infantry Regiment, commanded by Colonel Nelson, delayed the advance of the 116th Panzer Division for 36 hours. Even as enemy pressure increased, the 112th was in good shape, having destroyed 20 or so German tanks. Still effective, Cota wanted the 112th to retire to Bastogne, to bolster the defenses of the key road juncture. When movement to Bastogne by the 112th was deemed impossible, the unit joined the remnants of the 106th Infantry Division to the north and became part of the defense of St. Vith. Now Cota had just two infantry regiments to fight off two Panzer divisions.

On the right of the 28th, the 109th Infantry Regiment commanded by Colonel Rudder used the Sauer and Our rivers to slow the advance of both the 352nd Volksgrenadiers and the 5th Parachute Division. Four days later, the 109th was forced to withdraw south, joining the 9th Armored Division on the southern shoulder of the Bulge.

In the center of the 28th, the 110th Infantry Regiment commanded by Colonel Hurley Fuller took it on the chin, holding small unit position after small unit position, over 15 miles of rough terrain. There were numerous gaps in the line. Against the 2nd Panzer Division, supported by the 26th Volksgrenadiers, Fuller's infantry battalions and companies fought back, retreating from strongpoint to strongpoint. The 110th made its last stand in the medieval chateau of Clervaux, the "Alamo of the Ardennes," made famous by John McManus' superb book of the same name. Most importantly, when the order came down from Major General Troy

Middleton to General Cota to Colonel Fuller to hold at all costs, the men of the 110th did hold out.

Were the men of the 110th inspired by Cota's heroics at Omaha Beach, or the sacrifices of the men and women at the siege of the Alamo during the Texas Revolution of 1836? We will likely never know. The men refused to surrender, they refused to admit defeat, and they delayed the German advance. Fuller's 110th infantry regiment ceased to exist as an effective unit, reduced from 3,000 officers and men to 500. Its sacrifice gained crucial hours for the defenders of Bastogne.

With the fall of Clervaux, the Germans were over the Clerve River, heading for Bastogne with General Bayerlein's Panzer Lehr in the lead to get to Bastogne. They needed to capture Bastogne, and fast. The Germans knew that American Airborne troops, the 82nd and 101st, were heading their way. Inflicting horrific losses on the remaining elements of the 9th Armored Division's CCR, Panzer Lehr was pushing hard, driving what was left of Cota's 28th out of the way.

For the Americans, the defense of Bastogne proper had only just begun. The villagers of the town had been going about their business without interruption on December 16 and 17. That normality was soon shattered. Electricity failed, wounded and dispirited American soldiers began to flood in from the east, and refugees crowded the roads.

As elements of the battered 9th Armored Division CCR trickled in from the east, advance units of the 10th Armored Division began to arrive from Patton's Third Army. On the afternoon of December 18, Colonel William L. Roberts of the 10th Armored Division's CCB confirmed that Middleton would soon have three combat teams for the defense of Bastogne.

While Middleton was conferring with Roberts, Brig. Gen. Anthony C. McAuliffe walked in. McAuliffe's visit to Middleton's headquarters was purely social, to get the lay of the land. While

McAuliffe listened with interest to the events developing around Bastogne, his assignment was further north, or so he thought.

Hours earlier, First Army commander General Hodges realized there was little to stop Kampfgruppe Peiper in the event Peiper crossed the Ambleve River. He needed to create a defensive line west of the town of Werbomont. Gavin's 82nd Airborne Division was the closest unit available, closer than McAuliffe's 101st. Hodges ordered Gavin north to Werbomont and McAuliffe to Bastogne, the exact opposite destinations Gavin and McAuliffe had received from Eisenhower and Bradley. Gavin got the word and adjusted accordingly. On the other hand, McAuliffe had been out of radio contact.

When McAuliffe walked into Middleton's headquarters in Bastogne, Middleton thought McAuliffe was simply reporting as ordered. Instead, it was one of those fortuitous moments in the fog of war when the right person gets to the right place at the right time. Middleton briefed and updated McAuliffe on the current situation in Bastogne; orders went out to McAuliffe's G-3, Operations Officer Lt. Col. Harry W.O. Kinnard, to assemble the 101st in Bastogne. When Middleton departed, McAuliffe would be in command of the defense of Bastogne.

McAuliffe joined Middleton in conference with Colonel Roberts. Middleton wanted Roberts to disperse the combat teams of the 10th Armored Division into defensive positions, a use of tanks that defied armored doctrine. Commanders of armored units were taught the doctrine of concentration, not dispersal. Roberts pointed out that dispersal meant his forces could be chopped up piecemeal by German infantry with panzerfausts and anti-tank guns. Roberts added that dispersal meant there would be no opportunity to go on the offensive, to strike back at the enemy.

Middleton acknowledged Roberts' objections but pointed out that this battle was a defensive one. Roberts was ordered to deploy one combat team of the 10th to the east, to the village of Longvilly. Another was sent southeast, and the last to the north, to the village of Noville. Middleton then departed for Neufchateau, a town 15 miles to the southeast, site of VIII Corps headquarters. McAuliffe was now in command in Bastogne (see Map #2).

Roberts gave the Noville assignment to "Team Desobry," a force of about 440 men including a company of 17 Sherman tanks led by Major William R. Desobry. (The town of Noville, a picturesque village in a valley northeast of Bastogne, beyond Foy, was made famous in the series *Band of Brothers*.) Desobry got his men into defensive positions overnight. On the morning of the 19th, as darkness gave way to dawn, Team Desobry was confronted by fourteen panzers of Colonel von Lauchert's 2nd Panzer Division, arrayed in a skirmish line on the heights above the town. The morning fog had lain in the valley, concealing deployment of the panzers from American observers who could hear, but not see, the tanks on the ridge line. When the fog lifted, the Germans, reinforced by 50–60 other armored vehicles, attacked, opening fire on the Americans in the town below.

Desobry returned fire, giving as good as he got. In fact, the buildings of Noville provided cover for his tanks, and the newly arrived 609th Tank Destroyer Battalion destroyed nine of the 14 attacking panzers on the ridge above Noville. The defenders were able to hold their position. However, Desobry knew he could not repel a full-blooded assault. The Germans were in a position to pummel the village with artillery, to reduce it to dust, and to expose every one of Desobry's tanks. Desobry appealed to Roberts for permission to withdraw.

Roberts radioed Desobry that help, in the form of Lt. Col. James L. LaPrade's 1st Battalion of the 506th Parachute Infantry Regiment, was on its way. Throughout the day of December 19, LaPrade's men filled in the gaps, barely holding against increasing German pressure. As evening fell, Panzergrenadiers infiltrated the town, supported by panzers on the heights shelling the town. The fear that German shellfire would reduce the village to rubble and remove all cover from the defenders was coming true. Round after round exploded in the village, blocking the streets and leveling the stone buildings. Shell-fire eventually destroyed the second-story command post, killing LaPrade and seriously wounding Desobry. Major Robert J. Harwick took command of the paratroopers; Major Charles L. Hustead took command of the armor.

Reinforcement was attempted but failed. Paratroopers, like armored units, were not trained or equipped for defensive fights. When the 101st was abruptly ordered north, there was insufficient time to issue winter uniforms, camouflage helmet covers, heavy weapons, or extra ammunition. As depicted in *Band of Brothers*, the paratroopers scrounged weapons and ammunition from troops retreating from Bastogne; at the same time, they marched into the fight.

Brig. Gen. Gerald J. Higgins, the 101st's assistant commander, believed his men were "way out on a limb" and in danger of being wiped out. Colonel Robert F. Sink, the 506th's regimental commander, also wanted his battalion back. Back in Bastogne, McAuliffe believed it was time to withdraw from Noville. No withdrawal could be made without consultation with General Middleton. Having ordered Noville held, General Middleton would have to approve the retreat.

With assessments of General Higgins and Colonel Sink fresh in his mind, McAuliffe called VIII Corps headquarters. Middleton replied, "No. If we are to hold Bastogne, we cannot keep falling back."

The beleaguered men in Noville stood their posts, through the night fog of December 19 into the next morning. McAuliffe did what he could to reinforce Noville, sending forward the 705th Tank Destroyer Battalion, with its long-barreled 76mm guns, newly arrived from Simpson's Ninth Army. McAuliffe had Colonel Sink send a battalion of the 506th Parachute Infantry to Foy. The reinforcements were not enough to hold Noville, nor Foy. As dawn arrived on December 20, the Americans realized Panzergrenadiers had circled around and cut the road to Bastogne. Communications were lost, and the defenders of Noville and Foy were nearly surrounded.

The order was given: retreat to Bastogne. Throughout the day, the men of Team Desobry and the 1st Battalion of the 506th Parachute Infantry made their way down the gauntlet of fire. Losses were terrible; half the men, two-thirds of the tanks, and five of the tank destroyers were lost. The sacrifices were not in vain. The Germans lost 30 tanks and 600–800 men; the losses sustained in attack are almost always greater than among the defenders. More importantly, the blood and lives of the men of Team Desobry and the 506th bought time.

Bastogne was still in American hands, but soon McAuliffe and his men were surrounded. On the night of December 20, the road to Neufchateau was cut, effectively completing the encirclement of the village. Now alone, waiting for the promised relief from General Patton's Third Army, the men of the 101st resolved to fight to the last man. Stragglers were organized into ad hoc units. Ammunition was rationed; the artillery, down to 200 shells per artillery battalion, set a 10-round limit per gun per day. Food and medical supplies arrived by parachute.

The number of wounded men and civilians overwhelmed the field hospitals. The church in the center of town was transformed into a hospital. At first, the medics, nurses, and doctors of the 101st were able to keep up with the flow of the wounded. But as depicted in Band of Brothers, they were soon overwhelmed. Blankets on sawdust on the floor in front of the altar substituted for beds. Space was at a premium; even the altar was used as a surgical table.

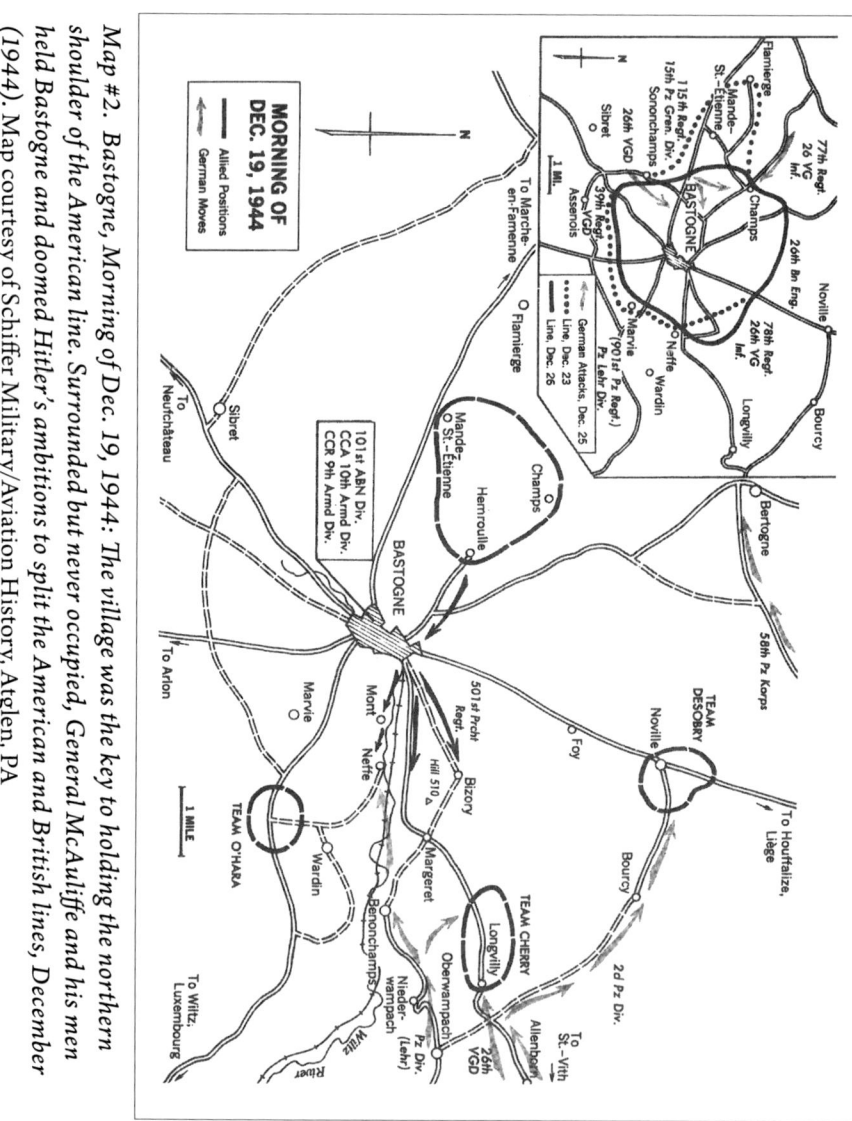

Map #2. Bastogne, Morning of Dec. 19, 1944: The village was the key to holding the northern shoulder of the American line. Surrounded but never occupied, General McAuliffe and his men held Bastogne and doomed Hitler's ambitions to split the American and British lines, December (1944). Map courtesy of Schiffer Military/Aviation History, Atglen, PA

While the 101st used the church as a hospital, the 10th Armored Division CCB took over the Sarma, once an elegant department store, for its wounded. A 30-year-old local Belgian nurse, Renée Lemaire, came to help out. In the autumn of 1944, when Bastogne was a quiet village behind the front lines, Lemaire met and fell in love with an American soldier named Jimmy. When his unit pulled out, Jimmy promised to return. To speed the end of the war and Jimmy's return, Lemaire volunteered her services to help in any way she could. She was a skillful nurse. Major John T. Prior, a surgeon, admired her work, but even more so, her presence. He said her presence "was a morale factor of the highest order." When silk parachutes from the airdrops turned up at the Sarma she asked Prior for one, so she could make a wedding dress.

In the meantime, the German noose around Bastogne tightened, to the point that the Germans thought the Americans would surrender. On December 22, four Germans—two officers and two enlisted men—emerged from the morning fog along the road to Arlon, south of Bastogne, under a large white flag. The senior German officer, Major Wagner of von Luttwitz' 47th Panzer Corps, was accompanied by Lt. Hellmuth Henke, of the Panzer Lehr Operations Section. The two enlisted men were panzer grenadiers. Americans in the area were members of the 327th Glider Infantry Regiment of the 101st Airborne Division. The German delegation stopped in front of the foxhole of a B.A.R. gunner, PFC Leo Palma.

The Germans had a message to present to the Commanding Officer in Bastogne. Speaking in English, Henke announced, "I want to see the commanding officer of this section." PFC Palma turned to Staff Sergeant Carl E. Dickinson, who motioned over PFC Ernest Premetz, a German-speaking medic in the platoon. They applied blindfolds to the German officers and led them in a roundabout way to the Command Post of F Company, a large foxhole. Captain James F. Adams, F Company Commander, read the message over the radio to the Battalion Command Post. As they waited, the message was passed up the chain of command to Major Alvin Jones, the Regimental Operations Officer, and then to Division Headquarters in Bastogne. Jones was ordered to Division Headquarters. The German officers were to remain in the F Company Command Post foxhole.

As Jones made his way to Division, Acting Chief of Staff of the 101st Lt. Col. Ned Moore entered McAuliffe's sleeping quarters adjacent to the Division Communications Section, rousted the general out of bed, and read him the contents of the German surrender demand. Half-asleep, General McAuliffe was in no mood for nonsense. His response, "Nuts."

Giving McAuliffe time to dress, Moore stepped out and briefed the Communications Section officers. By this time, Major Jones had arrived with the type-written surrender demand:

December 22, 1944

To the U.S.A. Commander of the encircled town of Bastogne:

The fortune of war is changing. This time the U.S.A. forces in and near Bastogne have been encircled by strong German armored units. More German armored units have crossed the river Ourthe near Ortheuville, have taken Marche and reached St. Hubert by passing through Hompre-Sibret-Tillet. Libramont is in German hands.

There is only one possibility to save the encircled U.S.A. troops from total annihilation: that is the honorable surrender of the encircled town. In order to think it over a term of two hours will be granted beginning with the presentation of this note.

If this proposal should be rejected one German Artillery Corps and six heavy A.A. Battalions are ready to annihilate the U.S.A. troops in and near Bastogne. The order for firing will be given immediately after this two hours' term.

All the serious civilian losses caused by this artillery fire would not correspond with the well-known American humanity.

—*The German Commander*

As Major Jones finished reading the message, General McAuliffe initially asked, "They want to surrender?"

Kinnard, the Division Operations Officer, responded. "No sir, they want *us* to surrender."

Taking the paper and reading the message himself, General McAuliffe said, "Us surrender? Aw, nuts!"

All agreed, a written response was called for. Kinnard suggested that the general's initial reply, "Nuts!" fit the bill.

The honor of presenting the response was reserved for the Commanding Officer of the 327th Glider Regiment, Colonel Bud Harper. Carrying the message to the Germans waiting at F Company Command Post, Harper informed Lt. Henke that he had a written response from the Commanding General in Bastogne. Still blindfolded, Henke asked what the response was. "Nuts!" he was informed. "Is that reply negative or positive?" he asked.

"The reply is decidedly not affirmative." Harper added this word of warning: "If you continue this foolish attack, your losses will be tremendous."

Lt. Henke and Major Wagner were then led back to the front. The small group was rejoined by the German-speaking PFC Premetz. Blindfolds were finally removed, and Lt. Henke read the written message:

> December 22, 1944
>
> To the German Commander,
> N.U.T.S.!
>
> —*The American Commander*

Still not comprehending the American slang, Lt. Henke asked again the meaning of the response. Colonel Harper was growing impatient and told PFC Premetz to say, "If you continue to attack, we will kill every goddamn German that tries to break into the city."

No longer feigning ignorance of English, Henke replied, "We will kill many Americans. This is war."

Harper added, "On your way, Bud, and good luck to you."

Understanding that negotiations were over, German Major Wagner saluted. Colonel Harper returned his salute.

Watching the Germans walk away, Harper called out, "If you don't know what I am talking about, simply go back to your commanding officer and tell him to just plain go to hell."

It was now 1400 hours. As the Germans departed, Harper fretted that his loss of temper might provoke a more determined assault. As it turns out, the threatened artillery barrage never occurred. The Germans lacked the transport capability to bring the required artillery shells forward. And increasing American air strikes were destroying German's ability to deploy field artillery in attack formations.

When Sixth Panzer Army Commanding General von Manteuffel learned of the surrender demand, he was furious. Furious, not because the demand was rejected, but furious because he knew he lacked the strength to carry out the threat to level Bastogne with artillery fire. Instead, von Manteuffel ordered the Luftwaffe to prepare a nighttime bombing of Bastogne.

And still the men of Bastogne waited for the promised relief from General Patton's Third Army. Messages like "Hugh [General Gaffey of the 4th Armored Division] is coming" received mid-morning of December 22 buoyed spirits for a short time. By the time Patton's message—"Xmas Eve present coming up. Hold on."—arrived on Christmas Eve, the defenders of Bastogne were in despair.

McAuliffe's reply to Patton, "Remember, only one shopping day left until Xmas," had a plaintive tone suggesting the defenders were growing increasingly weary. What was taking so long?

Patton had presented his plan for the relief of Bastogne on December 19. Approved by Eisenhower and Bradley at their meeting in Verdun, Patton promised an attack north to relieve Bastogne beginning on December 22. It was now Christmas Eve. Although Patton was moving north, for the hard-pressed men of Bastogne, surrounded by the 2nd Panzer Division to the north and Panzer

Lehr to the east and south, the counterattack was all too slow in coming. Part of the problem was disagreement between Patton and his old friend Middleton.

Patton came to Neufchateau on December 20 to confer face to face with Middleton on a plan of relief for Bastogne. During the conference, Middleton suggested Patton send the 4th Armored Division by the shortest route possible, the Neufchateau-Bastogne highway. Middleton knew the route to be less heavily defended by the Germans and the quickest way into the beleaguered town of Bastogne. Patton countered with a plan that required a wee bit longer route, the key difference being the use of the Arlon–Bastogne road. Patton argued that this route afforded greater opportunities to trap larger numbers of German troops in the Bulge, a pocket that now extended almost 50 miles into American lines. Patton conceded that his plan required the embattled troops in Bastogne to hold out longer, but he felt the chance to deal the Germans a fatal blow was worth the delay.

Patton wanted to go for broke and annihilate the three German armies in the Bulge, much like what had transpired in the Falaise Pocket in Normandy the previous summer. According to author Carlo d'Este, just two days earlier Patton had said about the German attack, "Hell, let's have the guts to let the sons of bitches go all the way to Paris. Then we'll really cut 'em up and chew 'em up."

Patton did attack on December 22, in the general direction of Bastogne, with the 4th Armored Division's CCA and CCB coming up the road from Arlon, the longer route. The attack was not a single thrust but a broad front attack, carried out by General Millikin's III Corps' 80th and 26th Infantry Divisions. Millikin pushed the Germans northward across country, an area with few roads and little strategic value. Millikin's right flank was covered by Major General Manton Eddy's XII Corps 5th Infantry Division under Major General S. Leroy Irwin. Patton hoped to create a noose around the base of the salient.

The defenders of Bastogne waited. As Leo Barron recounts in his book *No Silent Night (The Christmas Battle for Bastogne)*, McAuliffe rallied his men on Christmas Eve, saying, "What's merry about all this, you ask? We're fighting, it's cold, and we are not home. We men of Bastogne have stopped cold everything the enemy has thrown at us. We are writing a page in world history...giving our country and our loved ones at home a worthy Christmas present."

All along the perimeter, the men of Bastogne heard his words. They stood to their posts, shook hands all around, and wished each other Merry Christmas. They hoped and prayed General McAuliffe was right, a worthy Christmas present indeed.

General McAuliffe made his rounds, inspecting the troops and showing his face to the men he commanded in desperate circumstances. As he walked, he could hear the singing of "Stille Nacht" and "O Tannenbaum" coming from the town jail, holding German prisoners of war.

McAuliffe entered the jail to such taunts as, "We'll be in Antwerp soon!" and "We'll soon be freed and it is you who'll be the prisoner." Waiting for the Germans to calm down, and possibly recalling the 1914 Christmas truce that fell over No-Man's Land (a spontaneous cessation of hostilities along the Western Front with caroling, exchanges of food, and even a short England v. Germany soccer game), General McAuliffe wished them all a "Merry Christmas."

Then the Luftwaffe attacked, unloading a deadly cargo: two tons of bombs delivered by Junker-88 medium bombers. The tonnage was modest in comparison to the loads delivered nightly by the Royal Air Force (RAF) on German cities. But concentrated on a small Belgium village, the effect of the German bomber attack—as depicted vividly in the series *Band of Brothers*—was total devastation.

One casualty was the Belgian nurse, Renée Lemaire, who was working that evening in the Sarma. As the screech of the bomb increased, someone pushed her into the cellar, out of the kitchen where she was preparing supper for her patients. The bomb seemed to go straight down the chimney, erupting in the kitchen, killing everyone there. The building caught on fire and then collapsed, burying the dead and wounded, military and civilians. Lemaire's body was dug out of the cellar of the Sarma a few days later. Major Prior carried her body to her parents' home, wrapped in the white silk folds of the parachute that had been promised her for a wedding dress.

The Christmas Eve raid was preparatory for the all-out attack planned for Christmas morning, the Germans hoping to catch the Americans with their guard down. The 15th Panzer grenadier Division had arrived. They were fresh and well-prepared with white camouflage field coats. Two attacks were made. The most serious came from the northwest and penetrated as far as Hemrolle, a mile from the center of town. Like Peiper's men at Trois Ponts, they were stopped, their attack the high-water mark in the battle for Bastogne. German forces were exhausted, spent, or incapable of further offensive action. Bastogne held.

The next day, December 26, Lieutenant Charles P. Boggess of the 37th Tank Battalion, under Lt. Col. Creighton W. Abrams and supported by Company C of the 53rd Armored Infantry Battalion under Lt. Col. George L. Jacques, were welcomed to Bastogne by Second Lt. Duane J. Webster of the 326th Airborne Engineer Battalion. The siege of Bastogne was lifted. The butcher's bill was considerable: the 101st Airborne lost 1,641 men; the 10th Armored CCB lost 503 men; the 9th Armored Division CCR lost more than 700 men; and the 4th Armored Division about 1,400. German casualties are unknown but number in the thousands.

Until I began my research for *General in Command* and *To the Front*, I had no idea of the sacrifices made by ordinary men and women, Black and white, prep-school boys and college dropouts. I'd had hints in my young life of separation anxiety when I went off as a ten-year-old for eight weeks to Camp Wachusett in New Hampshire. My anxieties were real but could only hint at the profound pain and suffering of troops stationed overseas during wartime. As a child, I played soldier, running around the woods with Civil War swords and toy rifles. As an adolescent, I sang "Anchors Aweigh" and "The Caisson Song" at Army–Navy games, surrounded by West Point and Naval Academy veterans. For me, it was all fun, pomp and circumstance. I had no idea of the tragedy of war, of the lives lost in the snowy Ardennes Forest.

Now I realize my camp experience of sleeping in a cabin with 20 other boys and a counselor was a mild introduction to military life, just as Landon School Summer Day Camp and the Landon School were quasi-military institutions. When I later began to lose confidence in these places, I wandered on my own, reeling along without direction. Now I have come to realize their value. I was wrong to turn my back on the institutions that nurtured the men who won the Battle of the Bulge and who, like so many others of their generation, gave us the freedoms we now enjoy. But it would take me years before I realized the debts I owe. In the meantime, my adolescence continued.

12

Generals of the Battle of
the Bulge

The movie *Patton* was released in December 1969, to coincide with the 25th anniversary of the German attack. George C. Scott starred as the larger-than-life Patton, electrifying audiences around the world. Awarded an Academy Award as Best Actor, Scott refused to accept the Oscar, a move seen as a protest against the war in Vietnam.

Many of us were protesting the Vietnam War in our own ways. I had just turned 16. I had my driver's license and relished the prospect of wine, women, and song. Childhood dreams of glory as depicted in *Patton* were coming up against the reality of the *CBS Evening News* and film of combat in Vietnam. The Army officers on the nightly news seemed clownish, aping the lines fed them by politicians to justify our continued involvement in Vietnam. They seemed a far cry from the generals of the Battle of the Bulge. In my mind, no one with half a brain would join the Army, put themselves in harm's way, or get killed for America. No way.

And yet, deep down, I admired the nobility, the righteousness, and self-confidence of General Patton and others like him—men like my grandfather. They never talked about World War Two nor the role they played in the fight. They were either content to let history speak for them or they were so traumatized by the horrors of war—flamethrowers, high-altitude bombing, and concentration camps—they felt unable to justify their role in the war. Either way, back in the 1940s there was no *CBS Evening News* with Walter Cronkite with a microphone in your face, asking difficult questions. American generals of that generation were by and large apolitical and refused to answer questions better posed to politicians in Washington. In fact, Army Chief of Staff George Marshall made it a personal practice throughout his career never to vote in a presidential election. For him, it was Congress' responsibility to declare war; it was the military's responsibility to carry out orders.

In Britain, the line between politics and the military was often less well defined, especially in the European theater of operations in 1944–45. Whereas President Franklin Roosevelt had never been in combat, Prime Minister Winston Churchill had served on the front lines in World War One and frequently offered advice to his generals on military affairs. Whereas General Eisenhower claimed no interest in politics, Field Marshal Montgomery knew Winston Churchill shared his ambitions to lead the Allied coalition to victory over the Nazis.

At the height of the Battle of the Bulge, Montgomery demanded a change in command that would put him in overall command of all Allied ground forces, including the American First, Third, and Ninth Armies. Churchill supported this idea, at least in the beginning. It would add to British prestige, especially when the tide turned in favor of the Allies.

Montgomery's proposal was endorsed by his immediate superior Sir Lord Alanbrooke, Chief of the British Imperial War Staff. Both men had long resented the fact that under the current arrangement, General Eisenhower, Supreme Allied Commander, had British troops under his American command. It was not right that the colonists controlled the troops of the Crown. Like Montgomery, Alanbrooke had long sought opportunities to change the command structure and bring American troops under British command. Only seemed proper. So Alanbrooke took Montgomery's proposal to Prime Minister Winston Churchill. Believing that the crisis in the Ardennes allowed insufficient time to convene the Combined Chiefs of Staff, the group that would normally weigh in on important military and political matters that affected both England and the United States, Churchill breached protocol and took his concerns directly to Eisenhower. Late on the night of December 19, Churchill, ever the night owl, telephoned Eisenhower.

Churchill was unaware that Eisenhower had decided already to assign all American forces north of the line from Givet (in the west) to Prum (in the east) to Montgomery's 21st Army Group, to put American troops under British command, the first breach of the long-standing precedent established by General John Pershing in the First World War. Before Churchill could present his case that Montgomery should be given command of the American First and Ninth Armies, Eisenhower informed Churchill that both the U.S. Ninth and U.S. First Armies were being transferred temporarily to the 21st Army Group. Montgomery would be overall commander in the north; however, he would *not* be overall ground commander of all Allied ground forces. No, the change would only affect the American First and Ninth Armies. Bradley would retain Patton's Third Army. And Eisenhower would remain in command of both Bradley's 12th Army Group and Montgomery's 21st Army Group.

Churchill realized Eisenhower was not going to give up further control. Any efforts to convince him otherwise were likely to be harmful. Diplomatically, and deftly, Churchill responded, "British troops will always deem it an honor to enter the same battle as their American friends." Churchill was sensitive to Bradley's hurt feelings at the loss of command of the U.S. First and Ninth Armies. Montgomery would now be commanding more American soldiers than Bradley. Eisenhower's assurance to Bradley that the command change was only temporary was cold comfort; to the end of his life, Bradley said the command change was "one of the biggest mistakes of the war."

Eisenhower's response to the crisis of command was a masterful one. Any hope Hitler may have harbored for a significant split in the Allied command structure was thwarted. For Montgomery, the decision was also bittersweet. Even though he now commanded more men and had more resources than at any time in the war, four Allied armies and more than 30 divisions, he remained second-in-command to an officer he considered his inferior.

For Bradley, the loss of Eisenhower's confidence was a bitter pill. It was a sharp rebuke from his old West Point classmate and friend of thirty years. For George Patton, any concession to Montgomery was unwelcome; he could not understand making any compromise with the man he simply disliked, period.

Eisenhower's decision had consequences that reverberated along the entire western front but were felt most immediately down the chain of command to Hodges and Simpson. In the midst of the battle, Hodges now had to deal with Field Marshal Montgomery, always a critic of American generals. Hodges' reputation and the reputation of the men of the First Army were under assault, albeit less furious and deadly than the German attack. Hodges was doing the best he could, making life-and-death decisions with imperfect intelligence and with unreliable communications—decisions that affected the lives of thousands of American soldiers.

After Eisenhower ordered the change in command structure, Montgomery paid a call to Hodges' First Army headquarters. Upon his arrival, Hodges invited Montgomery to stay for lunch, a long-standing custom of the Americans that served as a gesture of goodwill and fellowship. Montgomery refused, saying he planned to eat alone, his box lunch and thermos of tea sufficient. The refusal was interpreted as evidence of Montgomery's inflated ego. But something else was going on; Montgomery was coming to First Army headquarters "like Christ come to cleanse the temple."

In fact, Montgomery wanted to relieve Hodges, whom he found distraught, haggard, and agitated, a candidate for a heart attack. Montgomery had no confidence that Hodges could carry on, especially in the dark days that likely lay ahead. Putting in a call to Eisenhower's headquarters, Montgomery wanted Eisenhower's approval for Hodges' relief. Eisenhower's response was quick, and diplomatic. "Hodges is a quiet, reticent type and does not appear as aggressive as he really is. Unless he becomes exhausted, he will always wage a good fight." Montgomery backed off, aware, as a British officer, that he would be hard pressed to relieve an American general at the height of the crisis.

We would do well to continue to keep our allies in any future wars apprised of our intentions and methods. President George H.W. Bush did so in the First Gulf War and removed Saddam Hussein and his forces from Kuwait quickly. The political goal of maintaining Kuwait's sovereignty was achieved. In the Second Gulf War, President George W. Bush was less forthcoming, and the political fallout was brutal. Unfortunately, in the recently ended Afghanistan War, the song remained the same.

Granddaddy's Role in the Battle of the Bulge

In addition to believing this history can serve as a guide today, my investment in preserving and sharing this history arises from a place of personal pride as I reflect on the role my grandfather played in the Battle of the Bulge. On December 16, Granddaddy and the XVI Corps were preparing to relieve the British XII Corps in the line at the northernmost shoulder of the Ninth Army sector. The German offensive, Wacht am Rhein (Watch on Rhine), interrupted their plan. When Lieutenant General Courtney Hodges' headquarters retreated from Spa, Belgium, he needed a new home. He found it in a fine building in Tongres, in east Belgium, recently occupied by Granddaddy's XVI Corps. Granddaddy's headquarters was now out on the street—and more importantly, he was out of a job.

When the remnants of the 106th were incorporated into General Matthew Ridgway's XVIII Airborne Corps, it was logical that General Alan W. Jones be designated Ridgway's assistant. Unfortunately, within days, Jones' heart gave out. Granddaddy was dispatched on December 26 by General Simpson to assist. Ridgway was known as a "hard charger," nicknamed "Old Iron Tits" for his habit of wearing a hand grenade on each chest strap of his web gear. The assignment was an awkward one. Granddaddy was senior to General Ridgway; in fact, Ridgway had been Granddaddy's pupil at the Command and General Staff College in 1935. Granddaddy's gentlemanly demeanor was not appreciated by the aggressive Ridgway, who likely felt his presence unnecessary. Granddaddy later wrote:

> … about the 26th of December, General Simpson called up and said Courtney Hodges had asked if he could borrow me for a few days, or at least until the Ardennes situation could be gotten in hand, the reason was that the commander of the 75th Division, whom I'll not name, wasn't doing so well, and he wanted to use me as division commander until he could recommend someone to take it permanently. I had had the 75th for only about two days, while at Tongres, and I too had been anxious about the commander. So, I reported to Matt Ridgway's corps, but Matt said he wanted to give this chap a chance as he thought he was doing better. However, I remained as a sort of assistant to Matt for about four days, but I could feel that Matt was somewhat embarrassed by my presence, as I was senior to him and also a corps commander. So, I told Matt that I was entirely available to him to use in any capacity, but if he felt the 75th was doing alright and he didn't want me to take it over, I thought I ought to return to my corps. My departure had started rumors that the Corps was being disbanded, morale dropped terribly, and I knew I had to get back to stop that. So, I called Courtney, told him the situation, and he said it would be alright to go back to

Heerlen. I told him I could return in half a day if he needed me. So, I got back to Heerlen about the end of the month, and morale jumped back to normal with the assurance that the Corps was not to be disbanded. The commander of the 75th did not last, however, and Ray Porter later took over the division. I have had it back with me from shortly after we crossed the Roer up to the present time, and it has done splendidly under Ray.

While Granddaddy was on assignment with the XVIII Corps, Hodges and Simpson moved more divisions from the Ninth Army to General Hodges' command: the 2nd Armored Division, the 84th Infantry Division, and the XVI Corps artillery. With only Granddaddy's former command (the 102nd Infantry Division, now under General Frank A. Keating's command), General Charles H. Gerhardt's 29th Infantry Division, and layers of artillery, Simpson held the northern extension of the right shoulder of the "Bulge," the Elsenborn Ridge.

As noted in Granddaddy's letter, the 75th Infantry Division was part of Ridgway's command. The 75th Infantry Division under Major General Fay B. Prickett was thrown into the battle despite having only arrived at the front on December 13, 1944. Originally assigned to Granddaddy's XVI Corps but diverted to the XVIII Corps for the Battle of the Bulge, the 75th found itself in a fierce battle alongside the paratroopers of the 509th Parachute Infantry Regiment under the command of Major Edmund Tomasik. In a bitter fight at Sadzot against elements of the German 2nd and 12th SS Panzer Divisions, the 75th recaptured the village using virtually just small arms and bazookas. Having helped as best he could, Granddaddy returned to XVI Corps headquarters, mission accomplished.

I knew nothing of this story until 2015. When I began reading my grandfather's letters, I learned the basics. When I began researching the subject, I found little in the historical record of my grandfather's service. When I wrote the Army War College, the library at West Point, and the researchers at the Command and General Staff

College at Fort Leavenworth, I heard back that Granddaddy was one of 34 corps commanders in the field. Beyond that bland fact, I was determined to fill in the historical record.

13

Roer River and Rhine River Crossings: The Dutch and the British

The Battle of the Bulge proved to be the last gasp of the once mighty Wehrmacht. With the destruction of three German armies, the tide turned. There was a 200-mile-long hole in the western front, through which the Americans and British began to pour, including Simpson's Ninth (see Map #3).

As one of the three corps commanders in Simpson's Army, General Anderson led the XVI Corps into the breach. For the next five months, from January to May 1945, one success followed another. Like a class president, captain of the football team, or a champion wrestler, Granddaddy enjoyed one victory after another (see Map #4).

The liberation of the town of Roermond, Netherlands, on March 1, 1945, was one such victory, notable for the actions of Sgt. Vincent A. Von Henke, a Polish-born, German-speaking scout for the U.S. 15th Cavalry. Von Henke was known to carry a Luger pistol in his shoulder holster as he infiltrated behind enemy lines, entering German-occupied towns and villages, drinking with the officers of the Third Reich, and learning the strengths and weaknesses of the German defenses. Under the cover of early morning darkness, he crossed the Roer River and entered Roermond, only to discover it devoid of Germans. Townspeople told him very excitedly that the Germans had realized they were outflanked and, threatened with being surrounded, they pulled out. Roermond's five-year ordeal of occupation, starvation, and desolation was over.

Notable in this saga are the efforts of African American soldiers under Granddaddy's command. One of my favorite photos from the time shows Granddaddy giving the order "Commence fire" to the Artillery Chief of Section, a sergeant who relays the order to the batteries standing by for orders. Granddaddy is in full field regalia: helmet, overcoat, .45 pistol in a holster under his arm, and swagger stick at the ready.

The Black artillerymen with their cannon are working hard, digging gun pits, and slinging artillery shells—the labor in the field familiar to all field artillerymen. While these men were working up a sweat, Granddaddy was traveling in an open-air jeep from gun position to gun position, inspecting the troops, and rallying them to the task at hand. Victory in World War Two required all the troops, from the lowest ranking to the commanders in the field, to give their best under circumstances few of us can even imagine.

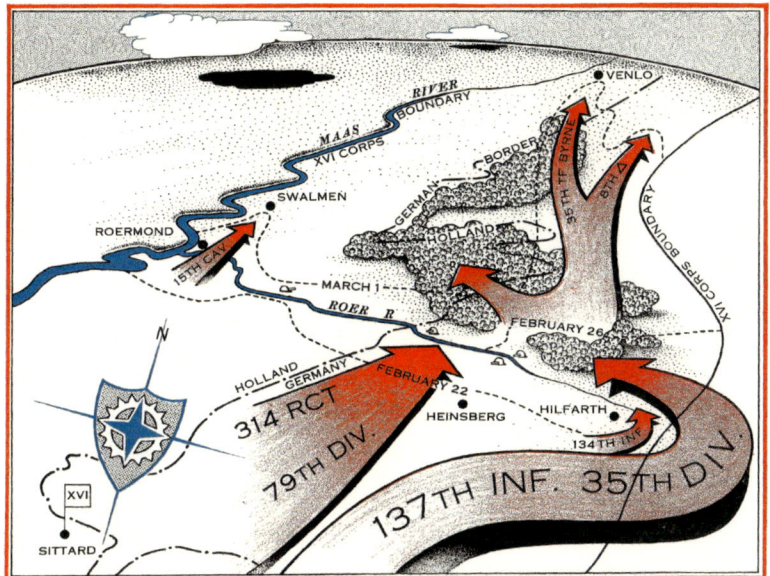

Map #3. Roermond and the Roer River Crossing by men of Major General Anderson's XVI Corps, February (1945). Schematic courtesy of Infantry Journal Press, Washington, D.C.

Map #4. Rhine River Approach by forces of the U.S. Ninth Army including Anderson's XVI Corps, February (1945). Courtesy of Casemate Publishers, Havertown, PA

The Roer River crossing was an important step forward. For nearly five months, the Germans had held firm, profiting from flood waters and bad weather to keep the Americans and British from advancing. Now with the Roer behind him, and with the weather improving, Anderson was feeling optimistic, as revealed in this letter to his wife:

> …We have had about five days of almost ideal spring weather. We are all enjoying it for more reasons than the mere fact that it is beautiful weather. Unquestionably, it is helping us to speed the final defeat of the krauts. The air efforts during these beautiful days have been tremendous, and, of course, it is also helping our ground efforts in the First, Third, and Seventh Army fronts. Georgie Patton certainly has been going to town again with his army. We do not know just what his score is, but unquestionably he has destroyed a nice sized section of the German army.
>
> Must close now as I have a busy day laid out.

Granddaddy sure did have a "busy day laid out." It was time to cross the Rhine River. The crossing on March 23–25, 1945, was weeks in the planning and was conceived to be a triumph for Montgomery and Churchill, a singular military feat with huge political benefit. Montgomery overlooked no detail of the set-piece battle, a battle plan designed to showcase his strengths. A huge artillery barrage to soften up the German defenses would prepare the way for an airborne assault. Once the paratroopers secured their objectives with isolation of German defenders on the east bank of the Rhine, Montgomery would then, and only then, unleash a massive amphibious attack over the Rhine. Montgomery would show the Americans and the world the brilliance of his generalship and justify Churchill's confidence in him. Artillery shells and bullets would spare the lives of his men, in contrast to the American's habit of relying on the initiative and bravery of individual soldiers or the Russian habit of squandering men in mass wave assaults against prepared positions. Careful preparation, overwhelming material support, and inspired generalship would highlight the strength of the Allied coalition, led by British and supported by the Americans, on a neat and tidy battlefield.

A week-long air attack of more than 10,000 Allied sorties preceded the attack, concentrating on Luftwaffe airfields and the railroad system. The goal was to isolate the battlefield and prevent the Germans from launching counterattacks during the early, most vulnerable stages of the battle.

The artillery phase of the battle, code-named Operation Flashpoint, was carried out by Anderson's XVI Corps. Simpson provided more than 2,500 artillery pieces of 105mm and 155mm caliber cannon from the Ninth Army for Anderson's use. Together, they fired more than a quarter million rounds over four hours. General Eisenhower observed the bombardment the night of March 23-24 from Anderson's headquarters at Lintfort, the arc of the shells lighting up the night sky. As author James Holland points out in his recent work *Brothers in Arms: One Legendary Tank Regiment's Bloody War from D-Day to VE-Day*, the Allies now had a huge material advantage over the Germans and used "steel not flesh" as much as possible.

Montgomery wanted an assault of three airborne divisions to secure the east bank of the Rhine. Code-named Operation Varsity, the plan was for U.S. General William "Bud" Miley's 17th Airborne Division to join the British 6th Airborne Division and secure landing zones within range of Allied artillery and link up with ground forces as soon as possible. After the usual rankling between American and British planners, General Ridgway's XVIII Airborne Corps assumed overall responsibility for planning the assault, under the ever-watchful eye of Montgomery (see Map #5).

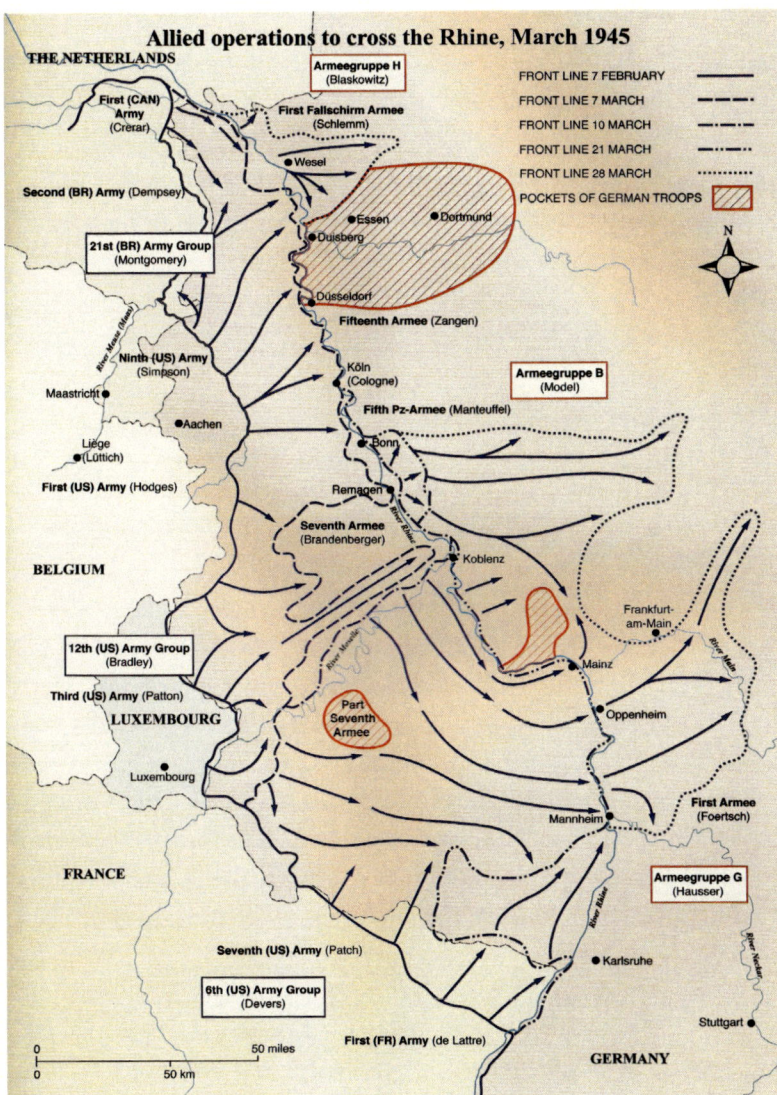

Map #5. Allied Operations to Cross the Rhine River, March 1945, included Montgomery's 21st Army Group in the north, Bradley's 12th Army Group in the middle, and Devers' 6th Army Group in the south. Map courtesy of Casemate Publishers, Havertown, PA

From Venlo, Netherlands, Winston Churchill observed the streams of paratrooper transport planes rumbling low over the west bank of the Rhine. Almost giddy with excitement, he longed to move closer to the battle, to see, and to be seen. With Field Marshal Alanbrooke in tow, after Palm Sunday services, he cajoled his way forward, first to Rees, then to Rheinberg, on the west bank of the Rhine. Churchill had full view of the pontoon bridges over the Rhine.

Granddaddy took time on March 25 to write:

> …Yesterday morning the XVI Corps crossed the Rhine. We were the only Ninth Army troops involved in the crossing, the remainder being British troops. Everything went off according to plan and we have a very substantial bridgehead on the other side now. I was on the east bank for a short while this afternoon when we took the Prime Minister across in a boat.

Winston Churchill crossed the Rhine River on March 25, despite Eisenhower's efforts to prevent it. By this time in the war, Eisenhower had seen Churchill in many settings, including formal high-level meetings at Tehran and Yalta, and informal moments of shooting Thompson submachine guns with the troops. Eisenhower knew Churchill's political instincts were powerful and that his propensity to hog the limelight trumped any sense of caution or self-restraint.

Granddaddy described the episode in a letter on April 6:

> No, Simp did not take the responsibility for letting Churchill cross the Rhine. The Old Man wanted to cross on the pontoon bridge, but I protested as the bridge was working at capacity and Churchill's party was so big it would add an unnecessary burden. So, I appealed to Simp and he in turn appealed to Ike, who said, "No!" After Ike and Bradley left, the Old Man saw one of our LCM's going across and asked if he could cross that way, and Marshal Montgomery said he saw no objection. That is the story of the crossing. The Old Man determined to get across, one way or another, however.

The photograph of the Rhine River crossing and the film taken of the event was a triumph for Prime Minister Winston Churchill. Published on the front page of every major newspaper in the western world, the image of Churchill surrounded by fighting men conveyed the impression that it was the British who were taking the fight to the Germans, the British who were winning the war in the west, and the Americans who were tagging along for the ride. Eisenhower, Bradley, and Patton were furious. Granddaddy and Simpson paid the price for failing to follow Eisenhower's lead. They were denied promotion, were denied postings in post-war Germany, and were shunned in post-war Washington.

Soon after the end of World War Two, the United States was called upon to defend the countries liberated by the Allies that were being threatened by the Soviet Union. Beyond assistance in the 1948–49 Berlin Airlift, our allies were content to join NATO and sit on the sidelines. In fact, France pulled out of NATO in 1966. Although the Cold War with the Soviet Union ended in the early 1990s, a series of undeclared wars around the world both during and after the Cold War has kept the United States busy. Called either "keeping the peace" or "police actions," these undeclared wars— in Korea, Vietnam, Panama, Grenada, Lebanon, Kuwait, Iraq, and Afghanistan—have sapped American resources and resolve for decades. The military leaders have rarely had the moral authority, much less the congressional mandate, for a declaration of war.

The framers of the Constitution divided war powers between Congress and the President. They made this division so that any decision to go to war would have the widest possible political consensus. Founding Father and future President James Madison pointed out that history demonstrates time and again the executive branch is the branch of government most interested in pursuing war strategies like the police action in Korea and Vietnam. In neither case did the United States declare war. While the President is Commander in Chief with authority to defend the United States, the Congress has the sole authority to declare war. More recently, Congress passed the 2001 Authorization for Use of Military Force (AUMF) and the 2002 Iraq AUMF to serve as the legal basis for military operations against suspected terrorist groups abroad. In practice, AUMF has meant the U.S. military going into harm's way without a full debate and vote on the merits of use of force.

The D-Day invasion of France and the liberation of Europe, called a "Crusade in Europe" by General Eisenhower, stirred the souls of both the public and the men and women in uniform. By way of contrast, the use of the word "crusade" by President George W. Bush to describe the Second Gulf War was roundly condemned, especially by our Arab friends who were offended by the connotation that Christians were coming to rescue them from Islamic sects hostile to their kingdoms.

American presidents are not alone in underappreciating the interplay of politics and military. During the Korean War, General MacArthur threatened to invade China and was relieved of his command by President Truman. In Vietnam, President Johnson's gradual escalation of our military commitment violated every tenet of military doctrine.

However, in the case of the Rhine River crossing in 1945, Eisenhower got it right. Churchill was looking to create the impression for the English populace that it was England that was winning the war in the west. For skeptics, read on.

In 2015 my wife, Sandy, and I traveled to London and visited the Churchill War Cabinet Rooms. At the end of the tour, visitors are led into a museum of Churchill's experiences in World War Two. Among the displays was a copy of the Rhine River crossing photograph. What a fitting tribute, I thought, until I looked more closely. The museum copy is cropped, excluding both Granddaddy and Simpson.

Unlike the Churchill War Cabinet photograph that focuses primarily on Prime Minister Churchill, Sandy and I found the people of the Netherlands grateful in their tributes to Americans in World War Two. A later encounter while traveling, not in Europe, but in Africa, brought this distinction home for me. We were in the Kalahari Desert on safari, and we gathered around the campfire at the end of the day and shared pictures of the animals we had spied. As the evening wore on, the darkness and cold of the night air deepened. We began to share stories of our own lives, our interests and passions. My story of the Rhine River crossing and the War Cabinet photograph seemed to evoke no more interest among the group than stories of sky diving in Delaware or snorkeling in the Virgin Islands. The next morning, we said our goodbyes to our safari friends. Before boarding the Land Rover, one member of the group pulled me aside. Since we were all going our separate ways, he knew this moment was his only chance for a private word. He had not wanted to say anything publicly.

Taking my hand and pulling me in close, the man told me he was a Dutchman and his family had survived five years of German occupation, humiliation, and near-starvation. He added that he was most grateful for my grandfather's service and efforts on behalf of his family and country. For a moment I was stunned, not sure of what to say, nor whether I could control my emotions. But I answered as graciously as possible that I thought my grandfather was glad to do his duty, contributing in a small way to the defeat of the Germans. I added that I believe he admired the fortitude of the Dutch and their resilience in the face of evil and cruelty. Then we parted. To this day, I am humbled that this man thanked me for the achievements of men like my grandfather.

In March 2020, right before the pandemic shut down most travel, I had a similar experience in the Netherlands. As mentioned earlier, my family was invited by city officials to attend the 75th anniversary celebration of Roermond's liberation by the 15th

Cavalry group, men under Granddaddy's command. They treated us like royalty, as though we were the heroes who had come into town under cover of darkness on the morning of March 1, 1945, not knowing who or what lay around the corner. The culmination of the three-day celebration was a military parade that ended in the town square—the same town square where crowds had gathered to see and hear Granddaddy address the crowd 75 years before. When I first saw the square, recognizing it from Army newsreel footage from 1945, I felt a shiver down my back and a weakness in my knees. Here was the spot of Granddaddy's only official honorary parade, given not by the U.S. Army, but by the citizens of Roermond. It was an honor to stand on the covered porch from which Granddaddy had received their cheers and delivered his victory speech.

Reading the letter he wrote to my grandmother about Roermond, I am not sure he understood at the time the significance of the event:

> ...On Wednesday, I have to go to Roermond, Holland, where at 2 o'clock a street is to be named for me— Andersonweg. Weg, I suppose, means street. It is in honor of the XVI Corps and is their method of expressing their appreciation for the liberation of the city by the corps on March 1st. In the meantime, I must prepare an address for delivery on the occasion. I was tactfully informed that it is always customary to mention the Queen and the royal family in any public address, so I've got to work that in somehow.

After the war, Granddaddy returned to Washington and a quiet life of family and friends. He arrived in the United States in a group of a dozen or so other general officers, none of whom had served together and known to each other only by reputation. They touched down in New York to little or no fanfare. Their aides departed and they scattered to their homes around the United States. For Granddaddy, the story of the end of the Trojan War may have come to mind. Ajax, the strongest and most heroic of the Greeks, had defeated the most valiant Trojans in one-to-one combat on the battlefield. Odysseus, the slyest of the Greeks, had tricked the Trojans into pulling the Trojan horse filled with Greek soldiers into the city, leading to its eventual destruction. Agamemnon, the Greek king, ruled from on high. When it came time to divide the spoils of war, Ajax found himself on the short end of the stick. He was furious, tearing off his armor, throwing his weapons away in despair. He believed Odysseus, the trickster, had outwitted him in Agamemnon's eyes and was rewarded with the lion's share of the booty.

I believe Granddaddy, and many others, felt like Ajax. They had carried the fight to the enemy and had won victory in the field. When the war was over, only a few among them were rewarded. I believe he was bitter at his treatment, and his bitterness led him to withdraw from all but his closest friends. And he drank too much. He kept in touch with many of the XVI Corps headquarters officers with annual reunions, usually around the time of his March 10 birthday. Despite many invitations from Roermond city officials, he never returned to Roermond. He never saw the reconstruction of the cathedral tower devastated by Allied bombing and artillery fire. He never saw the city come back to life, nor visited those citizens who saw him speak that day.

In 2020, when my family was invited to attend the 75th anniversary celebration of the liberation of Roermond, we jumped at the opportunity. For me, I had come a long way from the brash teenager, fearful of the draft and disdainful of General Patton as depicted in the movie *Patton*. I had come to serve active duty and retired military men and women who wanted nothing more than a kind word and a moment of my time.

14

Granddaddy's 75th Birthday

In March 1966, when I was 12 years old, I was invited to go with Granddaddy to The Army and Navy Club for the annual dinner of the XVI Corps Association. The occasion was to be a special one since the officers were using it to mark Granddaddy's 75th birthday. Dressed up in coat and tie, shoes shined, hair combed, I met him for the drive downtown to Farragut Square, near the White House. Granddaddy's car was a big, white Oldsmobile 98—yards of sheet metal and balloon tires that floated along. We arrived in grand style at the circular drive of the club, handing the keys to the valet like Rhett Butler handing over the reins of his horse at Tara.

The dinner was lavish; much of the toasting was lost on me but I could tell that Granddaddy was well respected, loved even, by the men in attendance. When presented a sterling silver plate engraved with the XVI Corps emblem and the inscription "From the Men and Officers of the United States Sixteenth Corps on the Occasion of your Seventy-fifth Birthday, with Affection and Pride," Grand-daddy was moved nearly to tears. His voice grew quiet and thick with emotion. He thanked the men, of course, and reminded them of the hardships they had endured together, not just the Battle of the Bulge, but those times when their future together was threatened by forces beyond their control. They stuck together during the boring days in Barneville, supporting the Ninth Army during the Normandy campaign before any combat troops were assigned to the Corps, and during the Battle of the Bulge, when he was assigned to the XVIII Airborne Corps to support General Ridgway in the counterattacks against the SS Panzer armies that threatened to tear apart the American line. They came together for the Roer River crossing, liberating the citizens of Roermond, with the 15th Cavalry leading the way. Days of glory.

He also thanked them for the near-Herculean effort of coordinating the Rhine River crossing with the British. He reminded them of the difficulty balancing the British demands for more men and more material with the needs of our own troops, and the glory-seeking commanders of some United States units. As he spoke, he acknowledged their smiles and nods, meeting the eyes of each one present. He knew it was unlikely, at their ages, that another opportunity to see one another would come along.

As he closed, he told one last story about the famous violinist Jascha Heifetz. Heifetz, a Lithuanian Jew, born in 1901, immigrated to the United States in 1917 to escape the purges of the intelligentsia by the backers of the Russian Revolution. (As an educated young man, Heifetz knew his days were likely numbered if he stayed.) He gained American citizenship in 1925. His playing was described and documented by Martianoff and Stern in their *Almanac of Russian Artists in America* in 1932 as: "an emotional experience, charged with beauty, rich with spiritual warmth…the great music of all time is interpreted through the medium of a faultless technique, illuminated by a rare and sensitive personality…one of the mysteries of great art—one of the secrets of Heifetz's art."

Here's my recollection of Granddaddy's story. He said, "When Heifetz arrived for lunch at our XVI Corps headquarters in Beckum, he was quiet and thoughtful. Of course, his command of English was superb, with just the trace of an accent. He played five selections, including his own arrangement of 'Deep River' and Prokofiev's *March from the Love of Three Oranges*. The men mobbed him afterwards, seeking autographs on German Reichsmarks."

Granddaddy went on. "A couple of weeks later, General Bradley hosted a dinner for the Russian general Konev, the Commander of the First Ukrainian Army Group. Bradley wasn't sure about the Russians, whether we should trust them or expect them to live up to their promises. He believed we should use every means possible to show our superiority. When Konev gave him a handsomely carved pistol, Bradley gave him a new American jeep, with an American carbine attached and the glove compartment stuffed with American cigarettes.

"Konev arranged a performance of the "Star-Spangled Banner" by the Red Army Chorus and a ballet by "a few girls from the Red Army." Bradley knew they were Bolshoi ballerinas.

"Since Bradley knew the Russians loved beautiful music, he asked Special Services in Paris to arrange a performance by a uniformed violinist. Konev was most impressed by the performance of what appeared to be an ordinary American GI Joe in Army fatigues.

"Bradley replied, 'Oh, that's nothing, nothing at all. Just one of our American soldiers.'

"Of course, the violinist was Jascha Heifetz."

Grandaddy then added another story about Heifitz.

"In a U.S. Army hospital in Italy in 1943, Heifetz played for a group of wounded soldiers. As Heifetz entered, those who could came to attention and applauded. One man who had lost his right arm waved his left hand in the air, beaming with delight. Heifetz was momentarily taken aback; nothing to date had prepared him for such a sight. After the concert, he spoke with his piano accompanist, Milton Kaye, and said about the one-armed soldier, 'You see, that is why we play.'

"You men here tonight remember, I am sure, the celebrities

and dignitaries that came to visit, to cheer us on in our efforts. None of us will forget Marlene Dietrich or Katherine Cornell. Such beauties, such grace. And those of you with me in Cannes on the French Riviera at the end of the war shed tears of joy and laughter as 'Sparky' Lang and Faulkner of the 102nd artillery, and Gatlin of the 406th Infantry sang the 102nd Infantry song for Bob Hope and Jerry Colona. Wonderful fun moments, but nothing moved me more than Jascha Heifetz."

Granddaddy handed me the silver plate. "What a beautiful tribute," I thought. And what a remarkable moment to share with him. At the time, I had no idea how important a moment it was, the last hurrah. Glory days with his old friends for the last time. The story came to mind recently as I read Evelyn Waugh's *Men at Arms*. In the book, the main character, Guy Crouchback, is conspiring with a friend to play a practical joke. His friend asks, "Why all the interest?" Crouchback replies, "When we are old men, memories of things like this will be our chief comfort."

And so it was with Granddaddy. Bradley's ploy to upstage his Russian guests amused everyone no end. It comforted them to know all the details. It was a story recounted for the last time in 1966. Shortly thereafter crippled by a stroke, Granddaddy never met with his XVI Corps staff again.

The silver plate was placed in the honor spot on a table in the foyer of his home, visible for all to see. It remained there, polished when tarnish built up, for the next ten years. Upon Granddaddy's death in 1976, Nanny put it away. She lived until 1991, and with her passing, the plate came to me. In the meantime, my own military service began in earnest.

In 1975, during my senior year at Yale, I needed to find a way to pay for medical school. Many medical schools, such as George-town and George Washington University, were charging $25,000 a year. My father did not have that kind of money. He made it clear that I would need to secure a scholarship to pay for school. The U.S. Navy had a scholarship program that paid all school expenses and a monthly stipend in exchange for military service, a year of service for each year of school.

It seemed fair to me. With a little push from my father behind the scenes—Chief of Naval Operations Admiral James L. Holloway was a classmate of his at the Naval Academy—I received the scholarship. Any misgivings I might have had about military service evaporated in the face of economic necessity. Some may read this account and say I had an unfair advantage, that I was a privileged white male who got a boost unavailable to others. I acknowledge that accusation, but would add that with privilege comes obligation. One cannot be separated from the other. With the scholarship came an obligation to serve one year for every year for which the Navy paid. My obligation was not reduced by training years; thus, internship (one year), residency (two years), and fellowship (two years) did not reduce the four years I owed. I wore the uniform a total of 13 years (four medical school, five years of training, and four pay-back years), subject to the rules and regulations of the U.S. Navy. It was a blank check I was glad to write.

15

Sea Voyages

I completed medical school at the University of Virginia in 1979. Following a one-year internship at Bethesda Naval Hospital, I was assigned as the medical officer of the USS *Sylvania* (AFS-2), a big supply ship stationed in Norfolk, Virginia. Along with two other supply ships bringing beans and bacon and the sundry items needed to check a Navy battle group on station, *Sylvania* supported operations of the U.S. Sixth Fleet in the Mediterranean Ocean. When I reported onboard, I had no idea of the hazards of shipboard life.

Sylvania was to be my home for at least twelve months, maybe more. She was a "McNamara wonder," hastily constructed in the mid-1960s as part of the former Ford executive's effort to bring the efficiency of civilian industry to U.S. Navy shipbuilding. *Sylvania* was part of the service fleet, not a true warship or ship of the line. Despite a pair of gun turrets on the foredeck, *Sylvania* was a thin-skinned and vulnerable tub of a vessel, riding high on the seas, ungainly, and squat in the beam. Not a handsome ship.

Officers and men who needed a second chance often found it aboard service ships, and many of the officers and sailors onboard were washouts from other Navy programs, like submarine or aviation programs for which they were ill-suited for one reason or another. *Sylvania* also served as a steppingstone for naval aviators who were

197

transitioning from flying to command of deep-draft vessels. If the captain, an aviator who took a year-long training program before taking command of a deep-draft vessel like *Sylvania*, did well, he would be in line to command a carrier, a nuclear-powered floating airfield and the cornerstone of the Navy's surface and aviation force.

After the captain and the executive officer, the senior supply officer was the ranking officer onboard. Because the delivery of supplies accurately and on time was highly valued by the captain's superior, and since every delivery was graded, the Supply Department on *Sylvania* called the tune. Delivery of supplies was accomplished while underway. In other words, ships on station did not return to base, or even stop cruising during replenishment operations. One part of underway replenishment was the delivery of pallets of food, clothing, and spare parts over steel cables stretched between two ships steaming side by side.

This method of replenishment had two great hazards. One was the possibility of a collision at sea. The helmsman of each ship steaming side by side had to maintain a true course with no deviation from the prescribed heading. Holding course was threatened constantly by the Venturi effect, the result of water squeezed and compressed between the two ships' hulls. The compression and pressure are greatest amidships. Aft of that point, the water accelerated, like water squirting out a hose nozzle. Water streaming faster, like the hose, results in decreased water pressure with the seeming paradoxical effect of pulling the two ships in toward one another, especially by the stern. The helmsmen described holding a true course as similar to dragging a truck and trailer with a flat tire; add another truck trailer in the adjacent lane and one gets a sense of the difficulty of the helmsmen's job. Notable is the hard truth that a collision at sea would likely be a career-ending event for a captain.

The other great hazard of underway replenishment was created by the steel cables strung from one ship to the other. The cables were high-tension lines, capable of suspending tons of valuable cargo above the swirling seas. In the event of loss of steerage, or a Venturi effect-induced collision, an emergency breakaway procedure would be initiated before the steel cables either pulled the ships into one another, or the cables snapped. Amputation, decapitation, and dismemberment by snapped steel cables were known to occur. Both a collision at sea and an emergency breakaway maneuver occurred during my time aboard *Sylvania*. Fortunately, no one was killed, or even injured, a tribute to the skill and expertise of the deck officers and bosun mates.

Sylvania also housed two CH-46 Sea Knight twin rotor helicopters. High lift capacity meant that the helicopters would complement the steel cable aspect of the underway replenishment effort. While *Sylvania* steamed alongside a destroyer or an aircraft carrier, steel cable loads going back and forth, the helicopters would lift pallets from *Sylvania*'s flight deck aft to the customer. A racetrack pattern kept the two helicopters separated from one another; round and round in a clockwise rotation they went, as quickly as possible, for during the time underway replenishment was occurring, the ships of the line were vulnerable. For example, no aircraft could take off or land on the carrier without interrupting *Sylvania*'s helicopter flight path. Believe it or not, on occasion, carrier flight operations did take place, grounding *Sylvania*'s helicopters for the duration of carrier flight operations. It was all a matter of priorities, all carefully coordinated, and all very dangerous.

Sylvania steamed independently, bringing needed materials, and occasionally men, to the Sixth Fleet Battle Group. During my tour of duty, *Sylvania* would frequently go back and forth to the Navy's Supply Center in Naples, Italy, load up, and then steam to the eastern Mediterranean to deliver to the battle group supporting the Marines in Beirut, Lebanon.

Sylvania would also make "house calls." On occasion, a frigate or destroyer with no medical officer would request medical assistance, not usually an emergency, but one never knew until after the fact. In those circumstances, since the "small boy" (I was learning quickly the jargon of the Navy where frigates and destroyers are called "small boys") had no flight deck, as *Sylvania's* medical officer I would go down in a horse-collar, a device used under the arms and suspended from a wire cable to lower a man from a hovering helicopter to the deck of ship steaming below, for a visit. Quite an adventure! I learned later that the rotors of a helicopter accumulate a significant electrical charge and a static jolt that can stun a man with the biggest static electric charge you can imagine. To prevent injury, a grounding rod is repeatedly touched to the leg of the person coming down from the helicopter overhead. Fortunately, I was never shocked. I was grateful to the sailor who carried out his mission to ensure my safety, though at the time, I was annoyed at his repeated pokes and smacks as I arrived on my mercy mission. Such is the life of a junior lieutenant medical officer newly arrived in the fleet.

After my first several months onboard, *Sylvania* departed Norfolk for six months in the Mediterranean. Up to that time, I had no idea what the deployment would hold for me. Although I was the medical officer, in truth I knew very little about keeping men alive, treating wounds, or preventing illness beyond CPR, defibrillation, suturing, and immunizations. I had only one year of practical training after medical school, namely internship. No intern does anything without a more experienced physician checking his work. An example of the unrealistic expectations of me was the captain's statement that should someone need an appendectomy, he would bring the ship into the wind to decrease her motion. He had no idea that I had never assisted in an appendectomy, much less performed one.

When all preparations that could be done had been done, it was time to leave for the Mediterranean. On the pier, wives and children of the officers and sailors living in the area waved enthusiastically. Kisses were blown, and children smiled and cried. I felt very alone, sad. I choked up and fled to my stateroom, not for a full-blown cry but in irritation and loneliness. I did not want any of my shipmates to see my quivering chin. A month or so before our departure, my mother and father visited the ship. My father was working at Langley Air Force Base for the NASA activity there near Norfolk, across Hampton Rhodes in Newport News, Virginia. Given that recent visit, there was really no reason for my parents to come and watch their youngest depart for a six-month deployment in peacetime to the Mediterranean, where the biggest threat to peace was Colonel Muammar Gaddafi of Libya. The resurgent Russian Navy had shot its bolt and its influence seemed to be waning, and the Palestine Liberation Organization was not likely to target a U.S. Navy supply ship. So I had discouraged my parents from seeing me off. I would have hated to show any emotion in front of them. I wanted to be stoic and firm. These were not my true emotions, but from my earliest years I was told not to cry because "big boys don't cry." From my perspective, it was better that my parents stayed away.

In retrospect, I wish I had better understood my own feelings. Yes, I was leaving for the unknown. Yes, I was on a ship, but I had been aboard long enough to know it was a second-rate ship with a second-rate crew. And yes, I was a gloried general practitioner, with few skills and even less experience. Leaving was a terribly lonely prospect. Then the whistle sounded, the heavy mooring lines were cast off, and the tugs sounded their horns. With a minimum of fuss, the tugs pulled us away from the pier, pushed *Sylvania* into the channel, and we were underway.

I am reminded now of the diary entry of Charles Tyng, a 13-year-old boy aboard a merchantman headed for China in 1815, which I read during a phase when I thought offshore sailing might be fun. In the first person, Charles writes:

> [I] listened with horror to the profane language of the sailors, who were under the influence of rum, and saw their disgusting looks and actions, and the idea that I was to be a

companion with such creatures perfectly shocked me. I had
no appetite to eat the dirty messes, and passed the night as
best I could, crying most of the time, overcome with such an
indescribable feeling of loneliness, and friendlessness, that
almost drove me crazy.

In contrast, there was no such sadness or temerity in Granddaddy's account of his departure from the United States in World War
One. His diary account reads as follows:

Originally, we were supposed to sail June 15, 1917, but
due to lack of transportation, or other reasons, we did not
receive our actual order to leave Douglas, Arizona, where my
regiment was then stationed, until July 21st, or thereabouts.
We were keen to go, and as our time was limited, we hustled
around, and all of us were clear of Douglas inside of two days.

We had excellent train service all the way to Hoboken,
reaching there in record time—about July 27th, I believe.
We were immediately escorted to the transport—*Henry R.
Mallory.*"

Mallory had been modified for military service with the removal
of all the second- and third-class staterooms and installation of berth-
ing for about 400 men. The officers were assigned the first-class staterooms. Gun platforms, kitchens, and toilet facilities were added for
the care of the men. Granddaddy and the Sixth Field Artillery were
on *Mallory* for her second transatlantic crossing, the destination of
the first crossing changed from Brest to Saint Nazaire by a submarine attack on the convoy. Granddaddy's diary entry continues:

On the morning of July 30th, at 2 a.m., we lifted our
anchor, and our convoy, consisting of our own ship the
Tenedores and *Pasadores*, carrying the 5th and 7th Field
Artillery respectively, a tanker, a cruiser, and three destroyers,
started on our voyage to France.

How he must have struggled to keep his emotions under control.
Pushing off from Hoboken, into the Hudson, down river from his
beloved West Point, in the middle of the night, gliding past the
Battery, a darkened Statue of Liberty, Sandy Hook, and the Atlantic

Highlands, past Breezy Point into the Atlantic. He calls the trip itself "quite pleasant and uneventful," without mention of cramped quarters, seasickness, stifling heat, high humidity, or inadequate ventilation. Maybe it was better in the first-class cabins reserved for officers like himself. Regardless, it would have been unseemly to complain.

No mention of homesickness, terror, fear, or anxiety, only:

> ...about two days off of France we were met by U.S. destroyers, whereupon our original protectors—the cruiser and three destroyers —turned about and started home. We were now in the danger zone, but nothing happened, except that we doubled our vigilance and our lookouts.

The only hint of anxiety is revealed:

> Finally, on August 12th, we saw land at about 8 a.m., and all of us were exceedingly thankful that we had come through without misfortune...scenery was beautiful, and it was hard to realize that this beautiful country was the scene of the most terrible, horrible and cruel war of history.

Not only did Granddaddy cross into the teeth of the German zone of unrestricted submarine warfare in 1917 with sang-froid and bravura, but he also repeated the trip in September 1944, departing New York for Firth of Clyde, Greenock, Scotland, on the *Queen Mary* with Prime Minister Winston Churchill aboard. Older and wiser, Granddaddy must have shivered at the thought of lurking U-boats and merchant marine ships sinking into the cold north Atlantic. No longer the second "Happy Time" celebrated by German Admiral Donitz in the spring and summer of 1942, the German submarine arm remained a threat. From the vantage point of 100 years after the fact, I am in awe of the strength of character, cheerfulness, and sense of mission reflected in Granddaddy's diary.

I hope Granddaddy and Smiling Jack took the opportunity to compare notes. My father was a member of the U.S. Naval Academy Class of 1943, which actually graduated in June 1942. Their time in Annapolis was truncated by the threat and then the reality of war. In

Michael M. Van Ness

anticipation of a need for naval officers, midshipmen stayed in school eleven months of the year from 1939 to 1942, foregoing the leisurely summer cruises typical of the peacetime academy experience.

During the war years, the Superintendent of the Naval Academy and the Brigade of Midshipmen hosted distinguished guests. In January 1942, in one of the most memorable of those visits, one attended by Smiling Jack, Lord Mountbatten, and an entourage of British naval officers lunched with the Brigade of Midshipmen in Bancroft Hall. Prominently displayed in Bancroft Hall is the "Don't Give Up the Ship" flag flown by Captain Oliver Hazard Perry to rally his forces, a flag that celebrates the United States' victory over the British Great Lakes Fleet in the War of 1812. Under that flag, Smiling Jack took his oath as a midshipman in July 1939.

Lord Mountbatten studiously ignored the banner; his job that day was to rally the midshipmen of the U.S. Naval Academy to the fight against Nazi Germany. In May 1941, shortly after *Bismarck* sank the "Pride of the Navy" HMS *Hood,* the rallying cry, "Sink the *Bismarck"* was heard everywhere, on both sides of the Atlantic, and certainly at the U.S. Naval Academy. When the British lost contact with *Bismarck,* all feared she would enter the Atlantic, roam up and down the Atlantic seaboard, and destroy all merchant shipping between the United States and England. "Sink the *Bismarck!* Sink the *Bismarck!* Sink the *Bismarck!*" It had to be done. And it was. The discovery of *Bismarck,* the rallying of British navy units from the Mediterranean, South Atlantic, and North Atlantic, like a pack of wolves intent on the kill, and the sinking of *Bismarck* electrified the hopes of the Allies during the darkest days of the war. Every midshipman knew the saga; every midshipman thrilled at the idea of one day being part of such a chase. My father told me that if Lord Mountbatten had asked for volunteers that day in Annapolis in the winter of 1942, he and his fellow midshipmen would have joined the British Navy without reservation or hesitation.

Smiling Jack did get his crack at the German Navy U-boats

roaming the North Atlantic. In June 1942, Ensign Van Ness was assigned to USS *Lansdale* (DD-426), homeported in Brooklyn, New York. As a young deck officer, he stood watches, straining to make out the periscopes of deadly U-boats, and armed depth charges dropped over the sides on suspected targets, 200 pounds of explosives detonated by a hydrostatic pistol actuated by water pressure at a pre-selected depth.

Smiling Jack also knew the reality of combat in the North Atlantic. The majority of casualties inflicted by the German U-boats were not directly from torpedo explosions, although they were horrific enough. The majority of deaths were from exposure, hypothermia from prolonged immersion in the icy waters of the North Atlantic. Smiling Jack completed nearly two years' service on *Lansdale*, protecting the invasion fleet during Operation Torch—the invasion of North Africa, protecting oilers carrying precious cargo from Aruba and Curacao to England, and shelling German panzers at Anzio on the Italian mainland. In early April 1944, Smiling Jack received orders to proceed to flight training, with the expectation that more naval aviators would be needed for the invasion of Japan in 1945.

On April 20, 1944, *Lansdale* was attacked and sunk by a glider bomb dropped by a Heinkel-111, one of a swarm attacking both port and starboard. "Abandon ship," was ordered by Lieutenant Commander D.M. Swift and carried out by his executive officer, Lieutenant Commander Robert Morgenthau, Jr., the son of President Roosevelt's Secretary of the Treasury. Survivors numbered 233 while 49 members of the crew either did not survive or were lost at sea. In 2010 my brother Scott and I visited Morgenthau in New York City. He was 91 years old, recently retired from the Justice Department as District Attorney for the Southern District of Manhattan, and generous with his time.

We spent most of the morning together in his office, sharing stories of World War Two. Although our father's service aboard

Lansdale overlapped six months with Morgenthau's time as executive officer, he had no specific memory of Smiling Jack. This was unsurprising to me, given the responsibilities imposed on the second-in-command of a destroyer in a war zone.

Scott recounted a visit he had arranged in 2005 with the pilot of the Heinkel-111. The pilot was living quietly in southern Germany with his wife. Welcomed in by the German, Scott reassured the pilot that no one was seeking revenge, only his side of the story. In broken English and lots of sign language, the pilot told his story of being shot down by *Lansdale* gunners immediately after delivering the glider bomb that broke the back of the ship. Expecting rough treatment from the Americans, the pilot was grateful, and amazed, to be tended to by the sailors and corpsmen who rescued him, all in accordance with the Geneva Convention.

Scott and I could have stayed all day and night with Morgenthau, but we knew we had already overstayed our welcome. Morgenthau had many people clamoring for his time—the lawyers in his office, the politicians with whom he had served over the years fighting organized crime in New York, and those who worked with him in his mayoral campaigns. Pictures of Mayor John Lindsay and Presidents Kennedy and Johnson were hanging along one wall. On the other three walls of the spacious office were Navy pictures of ships, the sea, and men in uniform. As we expressed our gratitude for the time he had given us, I remarked upon the numerous and prominent photos and artwork related to *Lansdale*. Morgenthau stopped and took a deep breath. "Men, those seven months on *Lansdale* made me who I am. No experience in my life was more powerful. When I took on the mafia, when I ran for office and lost, and when my wife died, I had the strength to carry on because of the Navy. That's why you see so much *Lansdale* memorabilia. I'll never forget those men."

Morgenthau's words make sense to me now, at this stage of life. Heretofore, I had never heard such sentiments. In fact, I heard very little from my father about his World War Two duty. He certainly held his memories of *Lansdale* close. He rarely offered stories about his twenty months aboard her. The fact that he attended only one *Lansdale* reunion in the late 1990s makes me wonder about his feelings about that part of his naval career. When he got back from the gathering, he had a reunion baseball cap and little else—certainly, no warm stories or memories of the days at sea. He only recognized a couple of the attendees and no one knew him. I think he was embarrassed to have wasted his time. Regardless, he never went to another reunion.

Years later, for his birthday, I presented him a framed print of *Lansdale*. He took out a magnifying glass and spent several minutes examining it in detail. Finally, he said, "She's riding high in the water, must be coming back from patrol. Gosh, I was glad to get off her and go to flight training." He had done his time; Navy ships are dirty, cold, wet, and noisy—and dangerous. He felt no survivor's guilt transferring off *Lansdale* within weeks of her destruction. Navy ships were targets and he had survived his time.

Lucky, too, was my father-in-law, Humphrey H. Cordes. In the late 1930s, he recognized the drumbeat of war and enlisted in the Navy, hoping to serve on submarines. When his aptitude for leadership was recognized, he received a reserve commission and ultimately commanded a cargo ship (LSM) transporting medium tanks toward Japan in preparation for an Allied invasion. At anchor near Tinian in the Mariana archipelago, he saw wave after wave of B-52 bombers setting off to attack Tokyo. One day, however, he noticed a rarer sight: the departure of just three B-29s toward an unknown destination. Only later did he realize that he had witnessed the *Enola Gay* taking off for Hiroshima, carrying the first atomic bomb.

Decades later, Cordes reported that news of the Japanese surrender August 15, 1945, had reduced him to tears of joy and relief. Like thousands of other servicemen steeling themselves for an invasion of the Japanese mainland, he had heard that Allied casualties were expected to exceed one million. Now he knew that he was likely to make it through the war unscathed.

While stationed in Japan after the war, Cordes toured the country and took special delight in visiting Kyoto, a pilgrimage destination for Buddhists. There, among countless Buddhist and Shinto temples, he happened upon a small Roman Catholic church. As a lifelong Catholic and a scholar of classical languages, Cordes was moved by this discovery and pleased to find that he could communicate with the Japanese priest he met there in a shared language: Latin.

Lt. (jg) Robert G. Wiltgen, the brother of Cordes' future wife, Rosemary Wiltgen, also served in the Pacific campaign. However, he was not as fortunate as Cordes. As a junior officer aboard USS *Twiggs* (DD-591), Wiltgen was already a veteran of the battles of Luzon and Iwo Jima when, in the late spring of 1945, *Twiggs* was ordered to Okinawa, the last steppingstone toward the invasion of Japan. Steaming towards Okinawa, the *Twiggs'* crew could see ship after ship being towed away from the raging battle, crippled by increasingly desperate and deadly kamikaze attacks. Upon her arrival off Okinawa, *Twiggs* was assigned to the Navy's outer defensive ring, the dreaded radar picket lines west of the island, organized to detect incoming kamikaze attacks.

There *Twiggs'* and Wiltgen's luck ran out. On June 16, 1945, a low-flying plane dropped a torpedo that hit *Twiggs'* port side, exploding her number two magazine. With its target nearly dead in the water, the plane circled back and crashed into the ship's bridge, Wiltgen's duty station. *Twiggs* sank in less than a minute, taking 152 men, including Wiltgen, to the bottom.

One final and noteworthy footnote to this story of Navy ships in World War Two—about the *Henry R. Mallory*, the converted Army transport ship that carried Granddaddy and the Sixth Field Artillery to France in 1917. When World War Two started, she was still afloat as part of the merchant marine force transporting men and material from the States to England. *Henry R. Mallory* was torpedoed and sunk on February 7, 1943, by U-402, part of a 20 U-boat wolf pack that attacked convoy SC-118, some 600 miles from Iceland. There were 383 passengers and crew aboard; among the 272 who perished that February day were the ship's master, 48 crewmen, 15 armed guards (U.S. Navy men who manned the 11 guns on deck), and 208 passengers (Army, Marine Corps, and Navy servicemen). SC -118 escorts included the USS *Schenck* (DD-159), a destroyer, and two Coast Guard cutters USCG *Bibb* and *Ingham*. Unable to stop the attack, their mission that day was to pick up as many survivors as possible, and to protect the remainder of the convoy. By way of revenge, USS *Schenck* is given credit for the sinking of U-645 in the North Atlantic in December 1943.

Three generations of my family experienced the uncertainty of separation from family, friends, and country. All but one of us were fortunate to return home to the United States, to resume our careers, and to be restored to our families. For some of us in the family, like me, the nostalgia of homecoming, the pain of returning experienced by warriors since the days of Odysseus and the Trojan Wars, was modest. For others, I am only now beginning to understand that some wounds are invisible and more pernicious.

The disease of alcoholism, the scourge of depression, and the tragedy of suicide are all too often the result. If we are to do justice to our returning service men and women, attention to their needs, derived from the service they have given, is needed. Granddaddy came home on an Army Air Transport with twelve other generals to be processed out of the service. No bands, no fanfare, and no official "thank you." Given what he had done, given what he had seen, and given what little time he had left in the service, it must have broken his heart. When he was decorated for his service three months later, the ceremony was attended by a photographer, his wife and daughter, and General and Mrs. Simpson. I am so grateful that General

Simpson made the effort, but where were the rest of them? Surely someone in the area could have made the effort. I know that Generals Bradley and Eisenhower were living in Washington in December 1945. Why were they absent?

Granddaddy's arrival reminds me of my return to the United States on a C-141 to a deserted air terminal in Philadelphia in 1981. No hotel reservation and no official greeting were not too surprising for a returning medical officer, but it would have been nice to have a Navy representative to answer a few questions. Left to my own devices, I arranged a midnight cab to my home in Washington, D.C.

Granddaddy's basement awards ceremony reminds me of my father's final Navy moments, when he received a Navy Commendation medal, the next-to-lowest decoration one can receive from the Navy. I was there, my mother was there, and a photographer was there. Someone from the Department of the Navy pinned the medal on my father's blouse and departed. Not much reward for 30 years of honorable service, for two-year convoy duty against the German wolfpacks stalking American merchantmen, and for flying F6F Hellcats and F8F Bearcats off of the decks of aircraft carriers.

The medal I received in 1982 for my 15 months of service on USS *Sylvania* ranked just below the Navy Commendation medal my father received. The ceremony was a fitting one; I am grateful that my mother and father attended. The Commanding Officer of Bethesda Naval Hospital did the honors.

As superficial as some may judge these types of award ceremonies and courtesies to be, I believe they go a long way in dignifying the service of our deserving veterans. It is the least we can do.

16

Sylvania (AFS-2)

As *Sylvania's* medical officer, I was in an unusual position. I had a professional, medical responsibility to the health and welfare of the men on board. I was the head of the Medical Department, with administrative responsibility for the corpsmen in the Department. I also was expected to manage a budget, a pharmacy, and an x-ray service. Periodic inspections by outside authorities looked for deficiencies on *Sylvania* in the U.S. Navy's hearing conservation program, asbestos screening program, and alcohol screening and rehabilitation program, called "Dry Dock." These activities and programs were to be set up, conducted, and supervised on board, utilizing traditional Navy management practices, including the time-honored chain of command. In other words, if I had a problem with the hearing conservation program that I could not handle, I was to report it to my immediate boss, the Executive Officer. If a sailor needed alcohol rehabilitation, again, I was expected to report that need to the Executive Officer.

I also had a professional responsibility to identify and report via a separate chain of command any medical problems that threatened the efficiency and proper function of the command. In other words, if I had a problem that could not be solved via the traditional chain of command, I reported the issue to the Fleet Medical Officer, a medical corps captain at Sixth Fleet Headquarters in Norfolk. Calling the Fleet Medical Officer was considered a drastic step.

Public attitudes toward alcohol use and abuse were changing rapidly in the post-Vietnam War era. Alcoholism was increasingly recognized as a disease, with both acute and chronic effects. Acute intoxication, alcohol poisoning, and alcoholic hepatitis were increasingly recognized as early signs of chronic alcoholism, a disease that afflicted all ranks, from the seaman recruit to the highest officer ranks.

Official policies reflected the shift in public attitudes. President Jimmy Carter and Chief of Naval Operations Elmo Zumwalt campaigned against the use of alcohol in any official setting. Whereas alcohol consumption onboard U.S. Navy ships was common as recently as the early 1970s, by the late 1970s alcohol consumption was strictly forbidden. A stiff drink in the wardroom after a hard day's work, forget it! A belt or two in the privacy of an aviator's stateroom to forget a close call, not allowed! A nightcap to speed sleep before tomorrow's hazardous duty, no! The use of alcohol for insomnia was no longer permitted.

The Medical Department did keep a store of "medicinal" whiskey for treatment of hypothermia, the whiskey bottles kept under lock and key, stacked in a safe (with the morphine syrettes reserved for battle wounds) and inventoried monthly by a representative of the Supply Department and the Medical Department. During my 16 months on *Sylvania,* neither the whiskey nor morphine was ever needed or used.

In the early summer of 1981, after being a member of the ship's company for over a year, we were in Naples, Italy, getting supplies for delivery to the Sixth Fleet Battle Group deployed off the coast of Lebanon. Although the United States was not technically at war, at times it felt like we were. Russian ships monitored our activities, at times even darting in and out of our path. Having grown up among Navy veterans of World War Two, I was in my element. Once when a Russian frigate ran a ship-to-ship missile up on the rail (an action analogous to loading a round in the chamber of a gun) and then pointed the missile in our direction, I was not surprised. In fact, I found the incident stimulating. Probably because no one got hurt.

The incident reminded me of a story my father told of guarding the beaches of Naples Harbor on shore patrol in early 1944. The harbor was full of Navy ships supporting the Army's efforts to drive the Germans out of Italy. The U.S. Navy was wary of saboteurs and forbad Italian fishermen, or anyone, from the beaches, especially at night. Enforcement of the ban rotated among the warships in the harbor. An officer armed with a 1911 .45 caliber automatic pistol led the ten-man patrol. Although the sailors carried heavier weapons, including Thompson submachine guns and M1 Garand rifles, few had ever shot them, and never in anger. A radioman accompanied the patrol. On this particular night, once ashore, the group fanned out and searched for saboteurs. Then they heard foreign voices in the distance and saw flickers of light. Charging forward with the men, ducking in and out, around the huge concrete anti-invasion blocks that littered the beach, my father soon found himself all alone. When the voices seemed to recede into the night, he stopped and thought, "What am I doing? This is nuts. The Italians are starving; they're desperate for food. They aren't out to sabotage our ships. Around the next corner, I might come face to face with one of these guys, a fisherman doing a little night fishing. What am I going to do then?"

My father beat a slow retreat, slow enough to avoid any charge of dereliction of duty, but quick enough to relieve his anxieties about confronting the unknown. He often told me, "I could have come face to face with a gang of desperate men. All I had was a .45 pistol. Luckily, I never got that duty again. Took too long for my heart to quit racing and my hands to quit shaking."

There I was in Naples, 35 years or so later. The anti-invasion blocks were gone. The Italians came and went as they pleased. In fact, every time we pulled into port, a welcoming committee of skiffs, powerboats, and pleasure craft streamed out to greet us and check out who was coming ashore. The Italians were frequently even welcomed aboard, such as seamstresses for repairing damaged clothing or tailors soliciting orders for custom suits.

Russian cruise ships were also present, tied up at the prime spots along the quay. It kind of gave me a creepy feeling to walk in uniform past the flag of the Soviet Union, flying freely from the stern, brilliant red with hammer and sickle, as Soviet sailors watched. I was sure the passengers, husky men and women, unsmiling and squinty-eyed, were there for more than a vacation cruise.

Naples had its beauty and charm. I was introduced to tiramisu, espresso, and real Italian pizza. Fresh mozzarella from the milk of water buffalo, Italian pepperoni (not meat, but yellow and red marinated peppers), and sparkling water—all wonderful and new to me. I never ventured off the ship for dinner alone; it was unwise to be in uniform, wearing a gold wedding band, and with money in your pocket, alone on the streets of Naples after dark. There was comfort and safety in numbers. Fortunately, the Executive Officer was assigned a car for his personal use. He liked to eat as much as the next guy. Adventuresome, but not overly so, he'd get a group together for supper; we'd sample the local wines, drink a little, but never return intoxicated—we were always on duty, since there was only one executive officer and only one doctor.

After one such evening, we returned, probably about 10 p.m. The Officer of the Deck informed me to "lay to the Captain's Cabin." This type of summons was common, but unusual at night; during normal business hours, I was often summoned via the one-mc, the shipboard loudspeaker system, "Medical Officer, report to the Captain's Cabin." Regardless of the time of day or night, I dropped whatever it was I was doing and reported. And so, I went there, with little inclination of the sequence of events about to unfold, the hair-raising ordeal about to begin.

As a bit of background: the Captain and I were well-known to one another. We had shared weekend rides to and from Washington D.C., drinking a beer or two along the way. We had stopped in Williamsburg, Virginia, so I could introduce him to my parents. Fellow naval officers and aviators, the Captain and my father shared stories of carrier landings and deployments aboard. Afterwards, I shared with them my concerns that the Captain was not what I had expected. He did not seem high-minded, or mission-driven, like my father and my father's Naval Academy colleagues. Too many times, it seemed to me, my Captain was cutting corners or using ship's resources for personal gain. There was a time when ship's electricians spent days working on the Captain's RV plagued with electrical malfunctions. This did not seem right to me, but I kept my mouth shut, except with my parents. They listened patiently; my mother offered up a platitude or two.

Early in my time aboard, the Captain and I would go drinking together. We shared pitchers of beer at Officer's Clubs in Norfolk, both of us reeling as we returned to ship. He helped me in my duties. When I uncovered inadequacies in the Medical Department, he was supportive of my efforts to improve its function, efforts that required difficult personnel issues and the arrest by the Naval Investigative Service of one of my corpsmen for drug dealing. And on a more personal level, the previous autumn, before our departure for the Mediterranean, he endorsed my application for advanced training in internal medicine. A few weeks later, when I was accepted to return to Bethesda Naval Hospital, he celebrated my success.

But strains remained, and worsened, especially once we departed for the Mediterranean and a six-month tour with the Sixth Fleet. During our time in Norfolk, I had grown increasingly concerned about the Captain's health and wellbeing. Initially my concerns about his health were vague. The amount of alcohol consumed by many of the officers, including the Captain, was substantial, but nothing beyond my own experience in college and medical school.

My alcohol use was also at times immoderate, leaving me hungover at work. I realized quickly that I did not need to go along to get along. In fact, exercise, reading, and writing were healthier for me, and making time for these activities instead of drinking ensured that I kept my professional distance and objectivity.

After nearly a year of close observation, it was obvious to me that the Captain's alcohol use was impairing his performance and the work of the ship. His behavior had become more erratic; odd accidents like diving into a swimming pool and cutting his face on a glass that was on the bottom of the pool, or falling off his gig while entertaining guests near the Isle of Capri, were becoming all too common. Some of these incidents were rumors, others witnessed by officers I trusted.

It seemed the best thing I could do, in the absence of receiving a cry for help or the occurrence of a major problem, was to distance myself from his inner circle. I chose to align myself more openly with the Executive Officer, a by-the-book commander with a forceful personality who was committed to improving the ship's performance. In Palermo, we played tennis together. With the Chaplain, we toured the Cathedral at Monreale. In Naples, we visited the ruins of Pompeii and drove the Amalfi coast with two other junior officers, U.S. Naval Academy and U.S. Merchant Marine Academy graduates. In the car and over dinner, we discussed our jobs and the trials and tribulations of shipboard life. Before long, we all expressed our concerns about *Sylvania,* how poorly she seemed to be functioning in her duties and what might be done to fix her problems.

The previous captain, executive officer, and crew all seemed to have gotten better results than we were achieving. *Sylvania* had received honors and decorations for her work, such as the "Battle E" painted on her smokestack for excellence in engineering and supply services. *Sylvania* had won the "Golden Anchor" award for the highest retention rate among her crew, meaning her anchors shone with gold paint for all to see and admire. *Sylvania* had been one squared-away, efficient, and capable ship. No longer. Low grades during recent inspections, low grades for accuracy of orders filled, and low retention marked her as second-rate. My friends were concerned. Although my future career path had been assured by acceptance into the Internal Medicine training program at Bethesda Naval Hospital, their futures were tied to the performance of *Sylvania*, the quality of which was increasingly erratic and unsatisfactory. The Navy meritocracy was unforgiving, dispassionate, and ruthless. Even if their officer fitness reports were glowing a 4.0 on a 1.0 to 4.0 scale, the reputation of *Sylvania* would handicap their promotions and future assignments.

That night, as I approached the Captain's Cabin, I was a little annoyed. Unless this summons was a medical emergency, could it not wait until tomorrow? I was tired; it had been a full day with a typical 18-hour day on deployment with routine tasks, sick call, and corpsman training. I put my irritation aside as best I could, knocked on the Captain's Cabin door, and entered.

The Captain was seated across the cabin on a sectional couch, in a flight suit, relaxing, obviously a little worse for the wear. Seated with him was the senior Supply Officer. The Captain asked me to join them in a drink. I hesitated.

"It's OK," he said. "It's outta my personal supply."

"No, sir. Thank you, no. I think I'll just hit the rack," I said finally.

"Ok, Doc. By the way, I wanted to ask you to perform my annual flight physical. Gotta keep that flight pay coming, you know. Can you do that?"

Again, I hesitated. A flight physical is an examination designed to ensure the health of naval aviators and their fitness to pilot aircraft. A flight physical is usually performed by a flight surgeon with specialized training in aviation medicine, a year-long course of study in Pensacola, Florida. I had not received that training and was not qualified to perform the exam.

"Well, maybe, since we're not near an aviation medicine facility," I replied. "Let me look at the form."

"Ok, it's around here somewhere. Come back tomorrow, after officer's call."

"Yes, sir, see you then," I answered backing out the door. "Good-night, sir." I nodded to the Supply Officer and departed.

I was qualified to perform a general physical examination and to draw lab work, just the routine blood count and metabolic function studies that show overall health, and I did so on the Captain. Because I had my suspicions about his health, in particular the effect of alcohol on liver, kidney, and gut function, my laboratory examination was comprehensive and revealing. Elevation of liver enzymes and depletion of magnesium, as well as increased size of red blood cells with a modest anemia confirmed my suspicions that alcohol use was threatening the Captain's health. And if the Captain's health were threatened, the mission of *Sylvania* was also at risk.

Now what? From a pure medical point of view, if I were in the Internal Medicine Clinic at Bethesda Naval Hospital, the answer to the question is easy. Sit the patient down, review the laboratory findings, and counsel alcohol abstinence. However, alcohol abstinence is often easier said than done. Risks include delirium tremens, aka DTs, a withdrawal reaction that carries significant risk of seizures, respiratory failure, and death. Would the Captain suffer DTs? The answer is hard to know, but daily alcohol use is one positive predictor of DTs. In other words, the alcoholic drinking daily is drinking to prevent DTs. Did the Captain drink daily? I didn't know, but the low serum magnesium level suggested that malabsorption of this essential mineral was impaired, a sign of significant alcohol use.

Fortunately, at the time, we were in Naples refilling the ship before heading back out to the Sixth Fleet. There was a Navy medical facility in Naples with laboratory, anesthesia, and operating room capabilities. They also had one of the Navy's Dry Docks, for detoxification and rehabilitation of alcoholics. I made an appointment with an alcohol counselor at the Dry Dock at the Navy Hospital Naples to discuss my Captain's situation, my suspicions, and findings.

After I presented his case, the alcohol counselor posed a series of questions. "Answer the questions as though you were the Captain."

"Ok," I replied, "I'll answer as though I were the Captain. But remember, he is on a fast track to command of an aircraft carrier."

She began the questioning: "How frequently do you drink alcohol daily? Do you drink to intoxication? Do you drink alone? Do you drink on board? Is the use of alcohol impairing your ability to carry out your duties? Do you want help to stop drinking?"

I said I believed the Captain would have answered "Rarely," or "No," or "Never" to each and every question—not because the Captain was an outright liar, but because I had learned during my internship that alcoholics often lack insight into the extent of the disease and its effects on their health, their family's health, and the health of others around them. In other words, alcoholics often minimize the effect their alcohol use has on them or others. I believe the Captain would have tried to protect himself, his career, and everything he had worked so long to achieve. If he had answered yes to any of the questions, his fast-track career would have been over. Despite the Navy's oft-repeated assurances and best efforts to treat recovering alcoholics no differently than someone with a chronic disease, alcoholism is a chronic disease characterized by frequent relapses for which there exists no guaranteed treatment.

I left there and concluded no help would come from the Naples Dry Dock. The discussion and conversations did help confirm in my mind that the responsibility rested firmly on my shoulders. I concluded that the Executive Officer was not in a position to take definitive action; the Captain was his immediate senior and was under no obligation to act on the Executive Officer's recommendation to seek help. In fact, for the Executive Officer to suggest the Captain seek help might have been considered a form of mutiny, or a power grab by an ambitious junior officer. I therefore concluded that my only recourse was to call the Fleet Medical Officer. I hoped he might give me some advice on the next step: Should I speak directly with the Captain? Should I discuss my concerns with the Executive Officer? Or should I wait and watch?

I needed to call the Fleet Medical Officer, but I couldn't do it from *Sylvania* without risking questions I was unprepared to answer. I needed to get back up to the hospital, about a ten-mile drive from fleet landing. There was no question of getting a cab or taking public transportation. I would need a car, and the cars were controlled by the Executive Officer (XO). I knocked and entered the XO's office. He was seated, reading a memo, making notes. Looking up, he said, "Hey, Doc, what can I do for you?"

"Sir, may I have one of the ship's cars to run up to the hospital?" I asked.

The XO stopped his writing and looked up. "Sure," he replied. "What's up?"

I really wanted to tell him the whole story, really wanted to seek his guidance and approval, before taking my next step. But I was conflicted. I thought I should protect him from what I thought was strictly a medical matter.

"I need to discuss a case with the folks at the Dry Dock," I said. "I need their advice."

"Anything I need to know about?"

I responded, "No." I think he knew I was dodging his question.

"You sure, Doc?" he asked.

"I don't feel I can talk about it. I'll let you know, OK?"

I could tell he was unhappy with my reply. Heretofore, we had shared our thoughts and feelings about problems, shipboard life, and fellow officers and men openly, without reservation. Usually, though not always, I would be the one listening to him; I would serve as a sounding board or counselor. On the few occasions that I had something to talk out, I felt comfortable laying it all out. So in this case, for me to hold back, his antennae began vibrating.

He paused, and then handed me the keys to his car. "See you back in an hour or two?"

"Yes, sir. That should be plenty of time. Thank you, sir. See you shortly." I turned and departed.

"What am I getting myself into?" I wondered.

The drive to Naples Regional Medical Center, the fancy name of the Navy hospital, was its usual adventure. In Naples, no one follows any rules. Traffic lights, lane markings, and speed limits existed to be ignored. If one were having a bad day and needed to work out some anger issues, driving in Naples would be therapeutic beyond all traditional remedies. Driving on the sidewalk, ignoring turn signals, and running red lights was exhilarating.

Usually I enjoyed driving in Naples. I liked the rough-and-tumble unwritten rules of the road. But not this day. Driving in Naples in a state of anxiety is hell. No one will yield for you to change lanes, the front edge of the bumper always has the right of way; if you want to change lanes, change lanes. Just be sure your bumper is ahead of the guy next to you. If a driver catches the eye of another driver so as to ask, "May I enter?" forget it. If you look, you lose. The best way to get into a line of traffic is to play chicken. Just stick your bumper ahead of the other car, and don't look. He'll yield. Despite these challenges, it was all too soon that I arrived at the hospital and sat down to discuss my plan with two Dry Dock counselors. I reviewed my concerns. I expressed my determination to speak with the Fleet Medical Officer. They offered a perfunctory question or two. A glance between the two of them. They clearly had discussed my dilemma before my return from *Sylvania*. I was led to a small office; the yeoman stood up and informed me how to use his phone, how to get a line to the States, and the number of the Fleet Medical Officer. Then the three of them departed, shutting the door gently.

I was on my own. With neither their endorsement nor objection, the counselors washed their hands of any direct involvement. The provision of a telephone with access to the States fulfilled their obligation. Making the call to Captain Phillips, the Sixth Fleet Medical Officer, was my decision.

I had met Captain Phillips the previous summer during the indoctrination course for general medical officers going to the fleet. He was a jocular, open-minded physician, a good teacher and sympathetic to the plight of young physicians with one year of post-medical school training under their belts heading out to sea. On most ships, there are no other doctors, no nurses, and no pharmacists. He knew it was a big responsibility to give to a physician who had never before practiced medicine without lots of support. Captain Phillips had made it clear to us that he welcomed our calls, whatever the matter.

I had worked with him the prior summer as I labored to correct problems I had identified in *Sylvania's* medical department. Those problems had been corrected, and my credibility was enhanced by the work we did together. As I dialed, I certainly hoped I was not making a mistake. My heart was pounding, and my palms were damp. I tried to remember the words of Sir Edmund Burke who said, "The only thing necessary for the triumph of evil is for good men to do nothing." I held fast to my conviction that I was trying to do the right thing.

The call was answered on the third ring. The duty corpsman informed Captain Phillips, who picked up immediately. Re-introducing myself and my duty station gave me a minute to calm down. My pressured speech slowed, and I poured out the details of the story. When I was done, I thought he might have a few questions. He did not. I thought he might offer some advice. He did not. I had hoped for some reassurance that I was acting properly. He gave me none. He gave no hint of approval or disapproval of my action in calling him.

He simply said, "Discuss this conversation with no one. You have done your duty. I will take it from here."

I thanked him for his time and hung up.

I returned to *Sylvania*. I returned the XO's keys. Then he asked, "Hey, Doc, you want to go play tennis?"

"Of course," I replied. "Meet on the quarterdeck in half an hour?"

He replied, "Sure, see you there."

As we drove to the tennis courts, near the hospital where I had called earlier, the XO studiously avoided posing questions about my visit there. We rallied back and forth. He was a good player, better than me. It irritated me that I could not beat him. I was younger, fitter, and more aggressive. But he was smarter. He used my strength to volley over my head or slice passing shots just out of my reach. We played about an hour, until the sun began to set. I was disappointed that I had not played better, that I had not given him a better game. And I said so.

"Don't worry, Doc. You got a lot going on," he said.

"Yes, I do." I responded.

"Wanna talk about it?" he asked. "Maybe it's just a tempest in a teapot."

He was fishing. I knew it and so did he. He knew me well enough to know I wanted to spill my guts, tell him the whole story, and relieve myself of at least some of the responsibility. But it was too late for that. My call to the Fleet Medical Officer set in motion a chain of events beyond my control.

Next morning, after breakfast, a grim-faced XO pulled me aside in the corridor. "Doc, you are to report without delay to the JAG office at the Supply Center. You've stirred up something lots of higher-ups want to know about."

I shouldn't have been surprised, though in truth I was. The JAG office at the Supply Center is the legal department covering *Sylvania*. I suppose I should have expected something like this; if I had discussed it with the XO, maybe I would not have been so alarmed. But I hadn't, so "Damn the torpedoes, full steam ahead!"

A Navy lawyer and clerk greeted me, took me into a small office, and swore me in: "Do you swear to tell the truth, the whole truth, and nothing but the truth?"

"I do, yessir," I swore. What else do you think I was going to say? In retrospect, I suppose I could have asked for my own legal representative, but I didn't think to ask. Anyway, since I had nothing to hide, just a story to tell, I pressed ahead, telling what I knew, what I saw, and what I had done. No more, no less. Then I was dismissed.

As I left the small office and entered the waiting room, I encountered not one but half a dozen *Sylvania* officers—young lieutenants like myself, and the XO, and the senior Supply Officer. He was really unhappy to see me and glared at me as if to say, "If looks could kill, you'd be dead."

"Gentlemen," I offered and got a couple cursory nods, tight smiles in return. I think they all knew I had started all this fuss, putting them in this awkward spot where their loyalty to their captain was to be tested against their loyalty to the Navy. Time for me to keep moving and return to the ship.

When I got back to *Sylvania*, I was greeted with an order to see the Captain. I knew this encounter would be painful and awkward. I suppose I hoped my call to the Fleet Medical Officer was going to absolve me of this moment. Nope, it did not. As is customary, I knocked, entered the Captain's cabin, and announced myself. "Captain, Dr. Van Ness."

The Captain was on the other side of the room. Turning quickly, with a flash of anger in his eyes, he growled. "Doc, do you know what you've done? You've ruined me, taken the wheels off my wagon; you know that, don't you? You have ruined my career."

"No, sir, I don't see it that way," I started. "My duty..."

"Knock it off, Doc. You have no idea of the damage you've done." Then he stopped; he had said what he wanted to say. He began picking through papers and clothing items, stuffing things in bags, packing roughly. Then quietly, seething, he added, "I am leaving in an hour. I have nothing more to say to you."

"Yes, sir," I replied. I turned and headed for my stateroom. I needed a moment to regain my composure. I felt sick to my stomach, nauseated and vulnerable. I thought to myself, "Have I done the right thing?" I believed I had, but maybe not in the right way. Maybe it was cowardly to have avoided the Captain. I didn't like to think of myself as cowardly. Wouldn't it have been braver, or more honorable, to have discussed my concerns with him before calling Washington?

After a few minutes, I headed to sickbay. Upon entering the space, my Senior Chief asked, "Hey, Dr. Van Ness, what's going on? I hear the Captain's leaving."

"Senior Chief, I think that's right."

"Really? What happened?"

I took a deep breath. It was not the time nor place to discuss my role in all this. "Chief, maybe later, OK? Right now, it's business as usual. Anything happening here, in sickbay?"

"No, Doc. Just the usual sailors' bumps and bruises after a night on the town in Naples. No VD this morning, either."

"Chief," I said, "that's the best news I've had all day. Got any coffee?"

"I do. No chief worth his salt doesn't have a pot going. You look like you could use a cup, or two."

"Thanks a lot, Chief. Thanks for noticing."

Shortly thereafter, the report over the one-mc announcing the Captain's departure rang out, "*Sylvania, departing!*" In the Navy, long-held customs like this one are cherished and honored in the performance. It may seem strange to use the name of the ship to denote the Captain, but it is Navy custom. And to confuse matters further, a ship is always a "she" and always called by her name, not "The *Sylvania*" but *Sylvania*. Thus, the announcement, "*Sylvania, departing!*"

I proceeded to the quarterdeck to witness the proceedings. I wasn't sure what I would say or do, but I thought it important to be there out of respect for the Captain's efforts. By the time I arrived, just a minute or two later, his car, a black four-door sedan, was pull-ing away. No last-minute recriminations or accusations. No banter or anger, just a car pulling away and a ship's crew wondering "What in the world just happened?"

And then the one-mc again. "*Sylvania,* arriving!"

"What? Didn't he just leave? What's going on?" I wondered.

What was going on was the arrival of the new captain. Usually a change-of-command ceremony is one of the most formal and scripted processes in the Navy. The outgoing skipper is honored for his service and the oncoming skipper receives advice and best wishes from all. That's how it usually goes, but not in this case.

The new captain strode up the gangway, small briefcase in hand, undoubtedly containing his orders, which he would present to the Executive Officer. Tall and fit, he gave a good first impression, turn-ing aft to salute the colors, and then holding his salute as he requested permission of the officer-of-the-deck to come aboard. He'd been informed that morning that he was to take command of *Sylvania* and put to sea as soon as possible, but no later than the following morn-ing. The pace of operations off the coast of Lebanon was increas-ing and demanding of food and re-supply. Any personal plans were canceled; they were secondary to the needs of the Navy.

Shortly, I was summoned to the Captain's cabin. I knocked and entered. The new captain was talking and moving about the cabin, stowing personal items here, official items there. His expression was one of irritation and preoccupation. Standing nearby was the Execu-tive Officer, with a sour expression. I had seen his expression before, all-business, wary, and stern.

"Sir, Dr. Van Ness, reporting as ordered," I said, trying to hide my anxiety. Hadn't I caused all this turmoil? I braced myself for a tongue lashing.

The new captain spoke, "I need some medicine."

He handed me a piece of paper. "Tagamet," I read aloud. "Yes, sir. We have it. I'll bring it up."

I thought, "A medicine for acid reflux; how appropriate. I hope Tagamet is enough to ease his heartburn. We should be so lucky."

Then, in a kind voice, the new captain spoke to the Executive Officer and me. "Let's get *Sylvania* back on track. Once the dust settles, I'd like to meet with a few of the junior officers, get to know them better."

And true to his word, a few days later, in Sicily, the Executive Officer got a small group of us together for dinner. One was a recent Naval Academy graduate, as was the new captain. That bond eased the conversation, as did a glass of wine. Even so, I was on edge. By going outside the usual chain of command, did they believe I had violated some unwritten tradition of loyalty and trust?

I think the new captain was also wary. Had he walked into a nightmare situation where he would be under unfair scrutiny? Were his officers teetotalers, out to make a name for themselves by reporting on their superiors? He needed to know the lay of the land, and what better way to get this knowledge than an informal get together away from prying eyes on *Sylvania?* As wine was poured, one question was answered—no, we were not teetotalers. We were not on some mission to bring back Prohibition. Stories of Naval Academy school days followed, including a few from my father's Naval Academy days. Career choices and future ambitions were shared. We were on the same page: we wanted *Sylvania* to excel. The evening ended on a toast to *Sylvania's* success.

Sylvania's performance did improve, seemingly overnight. To this day, I can recall no swifter turnaround of an organization than the one I witnessed on *Sylvania.* Although I am sad to this day about the first captain's relief, it was for the greater good. Yes, the wheels of his wagon were removed, his career path deviated. I am gratified he never commanded a U.S. Navy aircraft carrier.

For my part in *Sylvania's* history, I was decorated, receiving the Navy Achievement medal, an official recognition that a young medical officer made the right choice at a difficult time in a trying situation. Granddaddy's life of service, dedicated to the ideals of duty, honor, and country gave me the moral fiber to do what I judged I needed to do. From the perspective of 40 years later, I know I made the right call.

17

Paul "Lefty" Holmberg

As a child, I dreamed of being a Navy doctor at Bethesda Naval Hospital, a place I had visited many times in my youth for minor aches and pains, and for a knee operation at age 13. In medical school, I did two clerkships at Bethesda, short stints where the Navy and I sized each other up. Finding each other suitable, I applied and was accepted for internship at Bethesda beginning in 1979. Except for my year aboard *Sylvania,* I remained at Bethesda for nine years, first for additional training and then, beginning in 1985, as a staff gastroenterologist.

In the summer of 1985, I began an afternoon clinic of patients, expecting just the ordinary parade of retirees and military dependents. Having received no notice any VIPs, high-ranking active-duty officers, or congressional representatives, I was surprised by my last patient of the day, a tall, reserved, and distinguished-appearing gentleman about my father's age. In the Navy it is said that you have only one second to make a good first impression. This man made a good first impression.

His name was Paul A. "Lefty" Holmberg. He was a retired admiral, Naval Academy Class of 1939. Although wearing civilian clothes, he looked fit enough to still wear his dress blue uniform. On his lapel was a small, enameled bar, blue and white with a bronze star, denoting that he had received two Navy Crosses during his service. The Navy Cross is the highest award the Navy can bestow, just one rank below the Congressional Medal of Honor. And this man had two. Quickly I realized that I was in the presence of someone extraordinary. But he wasn't there for my benefit; he was there for my help because he was ill, very ill indeed.

He was in pain, his skin flushed red, and beads of sweat stood out on his forehead. Yes, Washington is hot and humid in the summer, but there was more to the story. In fact, my friend Dr. Bernie Powers had seen Holmberg the previous week and ordered tests. Since Powers had completed his time at Bethesda, Holmberg's follow-up care fell to me.

Holmberg had a carcinoid tumor, a low-grade malignancy that produces neuropeptides that cause cramps, gut ischemia, and even bowel infarction. When metastatic to the liver, as in Holmberg's case, carcinoid is incurable.

There were some things to try that could control the symptoms of pain and flushing. I prescribed an anti-spasmodic and an analgesic and asked Holmberg to return in a week or two to assess the effects of my palliative therapy. Having just recently met the man, I was not prepared to discuss prognosis.

As we were finishing up, Holmberg's eyes wandered to my bookcase where I kept some non-medical books I enjoyed, like *Miracle at Midway* by Gordon W. Prange. To my surprise, he said, "I know Professor Prange. Quite a writer." Without further explanation, he departed.

At the end of the day, it was common for me to unwind with my friend and fellow gastroenterologist Dr. Sarkis J. Chobanian. And so it happened that I told Chobanian about Admiral Holmberg, the Navy Cross lapel pin, and his offhand comment about Prange. "You'd better check that out," he said. "When Holmberg comes back, he'll want to know you looked into it."

And so, I did, only to discover that Holmberg was one of the heroes of the Battle of Midway. In June 1942, the Japanese mustered a huge fleet including four fleet carriers with the intention of taking Midway Island as a steppingstone to Hawaii. On his first combat mission as a naval aviator, a member of dive bomber squadron VB-3 flying from USS *Yorktown* (CV-5), Holmberg followed his squadron leader LCDR Maxwell Leslie in a steep dive on the Japanese carrier *Soryu* and scored a direct hit with a 500-pound bomb (see Map #6). He survived the bomb blast and the furious Japanese anti-aircraft fire to find *Yorktown* under attack and unable to retrieve aircraft. Out of gas, Holmberg ditched his plane on the fortuitously glassy sea, climbed out on the plane's wing, and with his rear gunner, hitched a ride in a lifeboat to the USS *Astoria*. Two months later, he flew against the Japanese again, achieving another documented bomb hit on a Japanese carrier.

When Admiral Holmberg returned to my office, we talked about his service, how he pressed the attack against the Japanese, diving at an angle of 70 degrees, steeper than the steepest modern rollercoaster, and how he despaired to find *Yorktown* listing to port, out of gas, and unable to land. He seemed gratified that I knew his story and appreciated the significance of his post-war achievements in missile and rocketry. In fact, he had worked with my father on ship-launched and submarine-launched missiles in the 1950s, work that was the foundation for intercontinental ballistic missiles carried to this day by the U.S. nuclear submarine force.

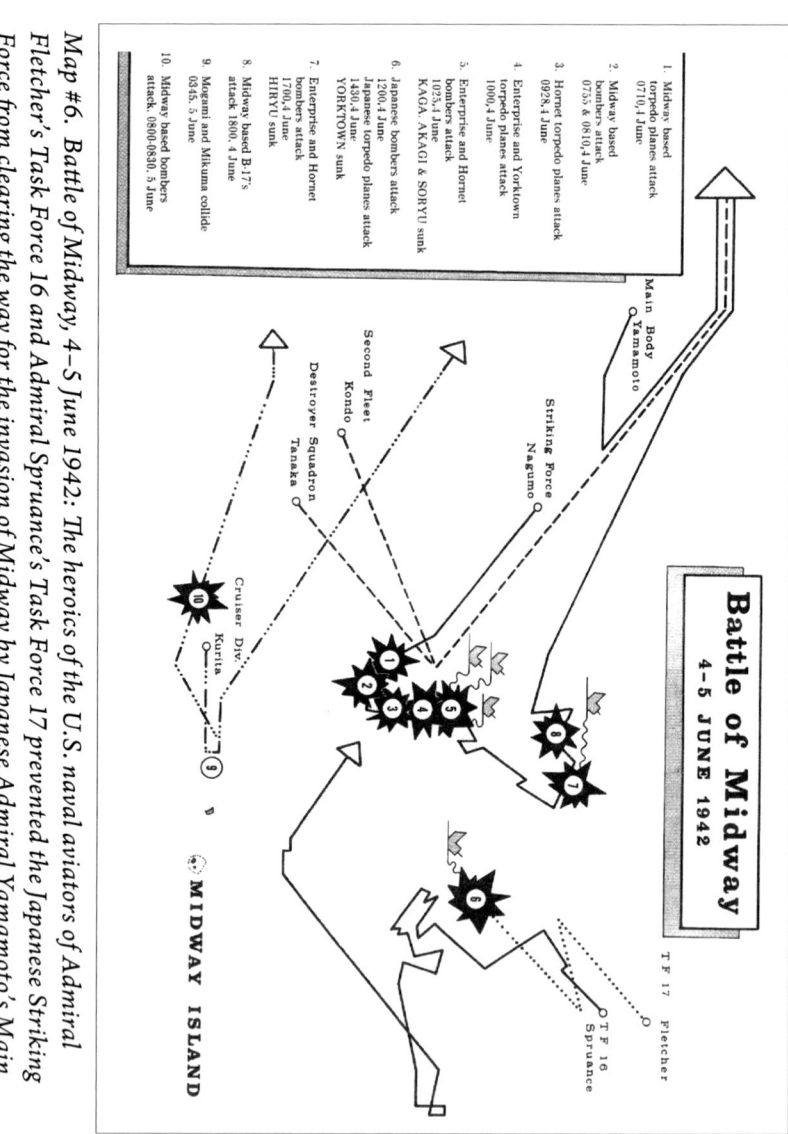

Map #6. Battle of Midway, 4–5 June 1942: The heroics of the U.S. naval aviators of Admiral Fletcher's Task Force 16 and Admiral Spruance's Task Force 17 prevented the Japanese Striking Force from clearing the way for the invasion of Midway by Japanese Admiral Yamamoto's Main Body, June (1942). Map courtesy of William Morrow & Company, New York, NY

I wanted to do something more to express my appreciation. At the time, the Vice President of the United States was George H.W. Bush. Bush had flown combat missions in the Pacific; in fact, he had been shot down and rescued by a patrolling U.S. submarine. So, I wrote the Vice President's office, thinking that he might be so kind as to drop Holmberg a note, one decorated naval aviator to another. Not sure that my letter would be taken seriously, I enclosed my copy of *Miracle at Midway* signed for me by Admiral Holmberg. At our next visit, Holmberg told me that Vice President Bush did write him a nice letter.

The last time I saw Holmberg, he was seated outside the cancer clinic, thinner and grayer that I remembered from the previous summer. Our eyes met for a moment, and I'm ashamed to say, I did not stop. I looked away. At the time, I had nothing to offer, no medication to ease his distress or to give him hope—or so I thought at the time. In retrospect, I did have something I could have given this giant among men. To sit with him for a minute or two, to acknowledge his suffering, to witness his distress and imminent demise, would have been a kind thing to do. I regret not knowing this at the time.

A recent visit to Arlington National Cemetery included a visit to Holmberg's gravesite, amongst the many heroes of World War Two, his friends and colleagues from his time in uniform. I placed a small rock on his tombstone to mark the occasion.

One small footnote: Dr. Chobanian's father, T5 John S. Chobanian, was a U.S. Army combat medic in World War Two, serving throughout North Africa and Sicily with his friend T5 John H. Malone. They shared the same food, the same duties, and the same tent for nearly three years before landing in Normandy as part of the June 6, 1944, D-Day invasion. On the night of June 11, Chobanian and Malone were settling down for the night. Malone went out to stretch his legs; after a long day in the battalion aid station, he needed a breath of fresh air.

The pre-D-Day bombardment of the area above Pointe-de-Hoc

and the surrounding area was so intense that the Germans pulled their artillery back from the heights. To guard their front, the Germans had sown hundreds of anti-personnel mines, the S-mine or "Bouncing Betty." American engineers had done their best to clear the ground, but were hampered by the uneven ground, severely cratered by American bombs and shellfire. Whether a misstep, confusion in the dark, or just a slip from a safe path, there was a muffled explosion. Malone was gone.

Malone was buried in the American Cemetery at Colleville-sur-Mer on the bluffs above Omaha Beach. Dr. Chobanian visited the cemetery in 2012, right after his father died. They had always shared a deep love of country and patriotism. It seemed a fitting tribute to his father, to pay his respects to his father's friend Malone. As Dr. Chobanian signed the visitor registry, he noted that his father, T5 Chobanian, had signed the registry, as well, in 1945.

We share a deep admiration for the heroes of World War Two, men who became our patients decades later at Bethesda. Men like then-Captain Daniel V. Gallery, commanding officer of Task Group 22.3 with USS *Guadalcanal* as flagship, who captured the German submarine U-505. Or Colonel John "Jack" Ripley who almost single-handedly held off a North Vietnamese tank battalion at the bridge at Dong Ha, receiving a Navy Cross for his courage under fire. Men like Marine Corps Colonel, then-Senator John Glenn who sought no special favor, making his appointments on time, checking in with the corpsman at the front desk, and waiting his turn quietly and without a fuss. How different his demeanor was from that of some elected officials who demanded appointments within 24 hours, expecting to be escorted from their cars by staff awaiting their arrival.

18

Wasyl Soduk

My military service ended in 1988. Although I could have stayed in and made the Navy a career, I decided to leave. It was not an easy decision. I gave up an important position as a specialist at the Navy's most prestigious hospital, where the most interesting and challenging cases were referred to internationally. I would no longer be a consultant to the Attending Physician to Congress. I would not follow in the footsteps of Rear Admiral Presley Marion Rixey, former Surgeon General of the Navy from 1902–1910, a distant relative on my father's side of the family.

I would have liked to stay in Washington, D.C. My parents lived in my childhood home in Bethesda, Maryland, my brothers were across the Potomac in Virginia, and many childhood friends and family still lived in the area. Unfortunately, the job market was terrible, saturated with newly minted gastroenterology graduates leaving the five training programs in the city. In order to get a good position with prospects for professional growth, I would have to leave. And so, I ended up in the Midwest, in Canton, Ohio, a great place to raise a family and a great place to start my own practice in gastroenterology.

One of my first patients was Wasyl Soduk, who overcame great adversity to come to Canton, where he lived gratefully and gracefully. Born in 1923 in a small village in the Ukraine, he is the father of three children, with five grandchildren and four great-grandchildren. It was not supposed to be this way. He was never expected to escape the slave labor camps of the Third Reich after being selected by village elders to fulfill a levy of workers demanded by the German authorities. He was supposed to die.

Wasyl was the second child of Constantine and Rusia. He had four sisters, one was older, the rest younger. He also had a younger brother, Mykola. Wasyl was a sickly child, the runt of the litter, a child who might carry his own weight but who would struggle to meet the demands of farm life. Whereas his sisters and brother were more robust, healthier, and livelier, his chores left him exhausted. He was thin, strikingly so.

The family lived in the small village of Monasteretch near the city of Stryi in current-day western Ukraine. The village has a difficult history, like much of the Ukraine, occupied by one foreign power after another. In the early twentieth century, Poland controlled the area. Wasyl's father served in the Polish army in World War One and Wasyl attended Polish schools as a youth. But soon, the Bolsheviks took political control of Ukraine, collectivizing the farms and exploiting the population for the "greater good" of Stalin's Marxist-Leninist regime. Stalin had made a pact with Germany, the Treaty of Rapallo, in 1922. In his book *Faustian Bargain,* author Ian Ona Johnson shows how extensive the cooperation was between Germany and Russia. The two post-World War One pariahs conspired to undermine the terms of the Treaty of Versailles. Germany received food and raw materials from Russia as well as space to develop new armaments of war. Stalin's Bolsheviks received advanced technology and training to modernize the Red Army.

With Stalin's ascent to power in the mid-1920s, the people of Ukraine lost Polish protection and were subjected to occupation, famine, and systematic plundering. As a Georgian, Stalin harbored no affection for the Ukrainians; in fact, he felt the exploitation of Ukraine was necessary for the Soviet Union's survival. The widespread confiscation of crops, farm animals, and farm machinery, along with the deliberate rejection of outside foreign food aid characterized the deliberate genocidal nature of Stalin's Soviet government to eliminate a Ukrainian independence movement. Before the Second World War, the worst depravity was the 1932–33 famine in Ukraine, called the *Holodomor,* the Bolshevik policy of collectivization which led to the confiscation of land, livestock, and crops from the farmers in the Ukraine. Collectivization caused widespread famine and the death of more than 3.5 million Ukrainians from disease and starvation.

Survival during the *Holodomor* was a moral and physical struggle. A woman physician wrote to a friend in June 1933 that she had not yet become a cannibal but was "not sure that I shall not be one by the time my letter reaches you." As described in excruciating detail by Timothy Snyder in *Bloodlands: Europe Between Hitler and Stalin,* the good people died first. Those who refused to steal or to prostitute themselves died. Those who gave food to others died. Those who refused to eat corpses died. Those who refused to kill their fellow man died. Parents who resisted cannibalism died before their children did.

All of this horror was hidden from the press and abetted by international communist sympathizers who argued that the larger goal of human progress was more important than the concerns of individuals, especially reactionary individuals who believed in God, freedom of speech, and capitalism. The writer Arthur Koestler wrote of these reactionaries, saying, "the starving are enemies of the people who prefer begging to work." Of course, he was referring to work for the Bolshevik state, labor for "the greater good." It seemed that many witnesses to the famine in the Soviet Union found a way to regard starvation not as a tragedy but as necessary for progress. Koestler avoided the truth about the famine, writing, "the destruction of the countryside could be reconciled to a general story of human progress." Koestler remained a true believer in Stalin's Bolshevik revolution and its communist ideals until the summer of 1939, when he saw swastikas flying from the Moscow airport welcoming Ribbentrop for the signing of the Molotov–Ribbentrop pact. The combined German–Soviet invasion of Poland sealed the fate of millions and confirmed to Koestler the mendacity of Stalin's socialist state. When Koestler complained to his housemate, the physicist Alexander Weissberg, that Ukrainians in Kharkov "have nothing to eat and therefore are dying like flies," he was told that everyone knew the truth, but to write about the famine would make their faith in Stalin and the communist dream impossible.

The reader may remember the expression from the Vietnam War, "We destroyed the village in order to save it." In truth, Peter Arnett wrote in 1968, "It became necessary to destroy the town to save it." Ralph Keyes is skeptical that Arnett quoted accurately the senior Army officer who always insisted that he said, "It was a shame the town was destroyed." Arnett's words echo those of James Reston, who wrote, "How do we win by military force without destroying what we are trying to save?" They are also reminiscent of the 1940 *Atlanta Daily World* editorial that the fight against fascism abroad was no excuse to accept intolerance at home: "We won't save democracy by killing it… and we won't make American democracy worth saving by destroying it in the so-called attempt to save it."

The radical French politician Edouard Herriot never denied the existence of the *Holodomor*. In the summer of 1933, after his third term as President of the Council and Minister of Foreign Affairs, Herriot visited Kiev. The famine was at its height, as were the Bolsheviks' efforts to deny its existence. The favorable testimony of a prominent French politician would further Stalin's efforts at international recognition. Herriot was the perfect choice; an apologist for the cause of international communism, he was also a Falstaffian fellow, a corpulent bon vivant enjoyed by all. He described his own body as "that of a woman pregnant with twins." After a 1933 visit to a Ukrainian orphanage, where children were coached to praise the care and feeding they received, Herriot traveled to a Moscow palace where he dined on caviar, all the time expressing his admiration for the communist revolution.

As a former French prime minister, Herriot knew the value of images and political speech, especially in the cause of a big lie. So did his hosts in Kiev. In preparation for Herriot's visit, the local authorities ordered the population of Kiev to clean up the city, decorating it with flowers and stocking the store windows with food and goods. Kiev became a modern-day "Potemkin village," a façade of prosperity, like the one Grigori Potemkin created to deceive Empress Catherine II during her journey to Crimea in 1787. Automobiles from surrounding cities were driven hither and yon by Communist Party officials, an illusion of a hustling and bustling city. At receptions, his hosts deliberately steered him away from Polish and German diplomats who knew the extent of the mass starvation. Herriot's hosts made sure there were no private conversations or back-channel communications; they would not spoil Herriot's fun.

In addition to keeping Herriot away from fellow diplomats, his hosts tried very hard to keep him away from the locals. When a woman on the streets of Kiev was able to get through and begged Herriot to tell the world the truth about the Bolshevik's mass starvation campaign, she was whisked away. Despite the starving woman's obvious distress, Herriot asked no questions. He did not want to deviate from the script—no need to ruin the party. Instead, he expressed repeatedly the lie that the Soviet Union was managing well to honor both the socialist spirit and Ukrainian national feeling.

Herriot saw what he wanted to see. Everything reinforced the lie, the lie of equity and prosperity for all under the leadership of the great "Uncle Joe" (Stalin). The children at the Feliks Dzierzynski Children's Commune in Kharkov, named for the founder of the Soviet secret police, were healthy, lively, and plump, dressed in clothes gathered for them the morning of the visit. The Stalinist dream of eradication of illiteracy and poverty was obviously successful to those whose eyes were blinded. Herriot wrote glowing words about the Soviet Ukraine; it was "like a garden in full bloom." In his book *The Black Book of Communism: Crimes, Terror, Repression,* Nicolas Werth recounts Herriot's words, "When one believes that the Ukraine is devastated by famine, allow me to shrug my shoulders."

Herriot might have fooled himself and gullible fellow travelers. Others, like German diplomats living in Kharkov and Kiev, knew better and reported to Berlin, saying, "Almost every time I venture into the streets, I see people collapsing from hunger." These diplomats also reported that many in Ukraine secretly hoped an invasion from abroad would relieve their agony. Anything had to be better than the Russians. In the summer of 1941, that invasion would occur, and many, including Wasyl Soduk and his family, were initially overjoyed. They would come to know that the realities of German occupation were not much better, especially when the Russians returned.

The German invasion of Russia in June 1941, code-named Operation Barbarossa, was conceived by Hitler as another blitzkrieg (or a lightning-strike war) that would bring vast amounts of territory, food, and wealth into the Third Reich. Had not the Netherlands, Belgium, and France capitulated before the might of the Wehrmacht? Many of Hitler's generals led by Field Marshal Goering believed similar results could be achieved from a bold strike to the east. Goering promised Hitler that a 90-day summer campaign would bring Poland, Lithuania, Latvia, Estonia, and the Ukraine into the Third Reich. Hitler is quoted as saying, "You only have to kick in the door and the whole rotten structure will come crashing down." No wonder Goering and Hitler believed they would capture Moscow, the grand prize that had eluded even Napoleon, who would soon become the second-greatest military dictator in western history. Food, oil, and labor were there for the taking.

These resources would create a huge, strong, inland empire capable of resisting British embargoes and blockades, of the sort that had strangled the German Empire in the First World War. In a nutshell, Hitler proposed to defeat the Russians, bring their riches into the Third Reich, and leave the Japanese to their own devices with their dreams of a "Greater Eastern Asia Co-Prosperity Sphere." It all depended on speed.

By August 1941, it appeared that Hitler's plans might work. Poland, the Balkan states, and parts of Ukraine were captured quickly and easily. All that remained of Hitler's plan for domination of Russia was the capture of Leningrad, Moscow, and Stalingrad. The riches of the Czar, held since 1917 by Stalin and his communist apparatchiks, and the Kremlin, seat of Soviet power, were only a few days of hard fighting away. In fact, the church spires of Moscow were visible through the binoculars of German officers. More importantly, the Caspian oil fields, and the German dream of oil independence were just over the horizon. One more determined effort, one last push by the mighty Wehrmacht, and Russia would be theirs!

Meanwhile, grand strategy was not the concern of the villagers of Monasteretch. Food, shelter, and survival were the day-to-day foci of the peasants and their families, who also worried about the unknown. What predations would the German army wreak on their little village?

For the first months of German occupation, daily life in Monasteretch went on as usual. The German army had run off or killed many of the ideologically motivated volunteers of the Bolsheviks who had befriended and betrayed so many in the countryside. Since the Germans lacked an intimate knowledge of village life, an experienced state police, and informers, they were unable to exploit the village as ruthlessly as had the Russians. For the German occupiers, it was best that the Ukrainian peasants continue their farming, the "surplus" to be reserved for the Third Reich. In fact, life under the Germans was better than life under Stalin—so much better that some in the village volunteered to go to Germany to assist in the war effort, either as soldiers or workers.

When the war in Russia began to drag on beyond the twelve

weeks initially estimated by Field Marshal Goering, more serious measures to exploit the Russian populace were needed, because the army could no longer expect to be supported by the civilian population of Germany. Rationing and privation at home would undermine civilian morale. Other measures had to be implemented. The German army was instructed to live off the land, like a conquering colonial power, and food was used as a weapon of war. With any so-called surplus food sent to Germany, starvation of the Russian population was deemed a necessity of war. Meanwhile, Hitler used Russia's refusal to sign the 1929 Geneva Convention as a pretense to relieve Germany of its obligations under the Convention to care for prisoners of war in a humane way. German soldiers were encouraged to consider surrendering Russian soldiers as sub-human cowards, not worthy of honorable treatment. Captured Russian soldiers were to be rounded up, caged, and killed. If not summarily executed, Russian prisoners of war were starved to death, as German soldiers were reminded that every bite taken by a Russian prisoner of war was a bite of food denied to a German baby.

Despite these draconian measures on the eastern front, all was not well in Germany; as the campaigns in the east dragged on, not only were food supplies decreasing, but the German war machine was starved for labor. Hitler's promises of plenty for the German population were unraveling. There were no unskilled workers for German farms—none for road building, wood cutting, or other public works projects. German youth were all gone, either in the army fighting overseas, in the hospital recuperating from wounds, or dead. And as the German army pressed forward (albeit more slowly than in the summer of 1941), more and more territory to control meant greater manpower shortages at home.

One solution to the manpower shortage were the youth of these occupied territories. They could not be trusted in conquered territory, but they could be used in labor camps established for them in Germany. In 1941 volunteer workers sent home reports of safe and humane conditions in the camps. These camps were not prisoner-of-war camps. Designed to keep the workers alive and working, there were rudimentary sanitary facilities, sleeping quarters, and medical clinics. Food supplies were simple and meager, but enough to sustain life. The German authorities recognized it would do them no good to work their labor force to death.

As the months wore on and conditions worsened for the German civilian population, however, so did conditions worsen in the volunteer labor camps. Food became scarcer, production quotas increased, and the laborers began to try to escape. More harsh conditions ensued. Rations were restricted further, meaning more died of disease. And punishments became more brutal. In the Hamburg munitions factory where Wasyl Soduk's future wife, Anna, toiled with hundreds of other women, workers' complaints were met by beatings and rape. Nighttime bombings by the Royal Air Force (RAF) began and terrorized everyone. Suicide was common. The number of labor volunteers and the quality of their work in the vital munitions industry dropped dramatically; they were no longer available in numbers sufficient to keep the German war machine running. By early 1942, the camps were not operating efficiently enough to meet the production goals of the Reich.

In the spring of 1942, the mayor of Monasteretch was informed by the local German military authorities that he was to select men from the village for labor duty in Germany. Making the mayor complicit in the process absolved the Germans of any direct responsibility. Once identified, the men would be taken in the middle of the night, preventing anyone from doing anything about it. And so it was that in the middle of the night, in June of 1942, Wasyl Soduk, aged 19, was rousted out of bed. He and his family had learned that he had been selected, but since there was nowhere to go and no way to avoid the inevitable, his mother had prepared for this moment by creating a small parcel of goodies. The bag contained a loaf of green bread—green, as the summer wheat had not yet fully ripened.

The wheat's nutritional value was incomplete, but it was better than nothing. Within minutes of the loud banging on the door and the shouted commands of the authorities, Wasyl was gone, never knowing if he'd see his parents or his home in Monasteretch again. His mother stood in the doorway watching the truck speed away, taillights fading into the dark night.

On the way to Germany, the train made one of many stops along the way, at a railroad junction in Poland. At most stops (on sidings, to allow other trains to pass), the men were kept locked in the stifling cars, called "vagonu," poorly ventilated and without sanitation facilities. On rare occasions, especially out in the countryside, the men were allowed to disembark, stretch their legs, and breath some fresh air. This stop in southern Poland was different. The guards, going car by car, heaved open the heavy sliding doors. The bright sunlight blinded the men. Given only a moment, the men were herded into an enclosed area. In addition to the train guards, the men noticed many additional soldiers, weapons at the ready. Surrounding the yard, giant German shepherds were held at bay by their handlers. The men knew the stories of the Jews of Lvov murdered by viciously trained attack dogs and were especially frightened by them. Staying clear of the animals, the men obeyed the guards' instructions without question or delay.

Wasyl and the other men were ordered to strip naked. A thorough examination followed. Wasyl was 5 feet tall, weighed only 110 pounds, and was uncircumcised. The uncircumcised men were separated from the circumcised. Germany was being "cleansed" of Jews, and since the circumcised men were presumed to be Jewish, they were not going to Germany. Instead, they were pushed along to another siding, where they were ordered to board other closed cars. They could not resist. The guards were too vigilant and held rifles and pistols at the ready. The dogs were too close. Once they were loaded into railroad cars and the sliding doors shut, the men's fate was sealed. The railway line led north, ending at Auschwitz.

With the slamming of doors and the cries of their fellow travelers fading in the distance, the uncircumcised men were deloused with DDT and herded onto open-air railroad flatbeds. They were each issued a filthy, thin, wool blanket and shown a hole in the floor for defecation. Hard to fathom, but this change was an improvement in their travel conditions. Wasyl huddled with his fellow conscripts as the train headed west, destination unknown. The trip was three days of torture, with the cold night air chilling him to the bone. From this point on, stops were few and far between. Wasyl realized his small packet of goodies with the loaf of green bread and other goodies was gone, stolen sometime in the night as he slept. Its loss was the first of many hard knocks. Minimal food, thin soup and crusts of black bread, and water were provided. Soldiers, responsible for the delivery to the labor camps of the exact number of men loaded in Ukraine, monitored their every move.

Arriving in southern Germany, likely the Black Forest near Munich, Wasyl began to work. As a laborer, he was judged a weakling, unlikely to last long; but as one of many peasant laborers, he likely posed no threat to the Reich. Keeping quiet and offering no resistance to orders meant survival. Under the pretense that the labor camps were simply part of the war effort, Wasyl was allowed to send and receive mail from home. In a letter in 1944, he learned that the Russians were on the brink of driving the Germans out of Monasteretch. Then, when the Russians drove the Germans out, the letters stopped. Wasyl was to learn later that the return of the Russians meant the conscription of his brother, Mykola, into the Russian Army.

Wasyl survived the next three years, and in fact thrived in the Black Forest, cutting trees for timber to shore up the mineshafts of the area coal mines. He avoided trouble at every turn and even dodged the machine-gun bullets of strafing Allied aircraft who mistook the lumberjacks for German soldiers. As the war entered its third year, German fortunes declined under the weight of the

Allied air campaign. The need for timber decreased, and Wasyl and his Polish friend Josef were released from the labor camp to work on a German farm near Landstuhl. The farmer was off at war, as were his three sons. The farmer's wife and her teenage daughter were doing their best to hold the farm together, but it was a struggle. Wasyl was welcomed on the farm, slept in the attic, and was treated much like a member of the family.

Then, Wasyl fell ill with a mysterious and severe stomach pain that doubled him over. The farmer's wife knew Wasyl to be a stoic man. Alarmed by the severity of his symptoms, the farmer's wife arranged for him to be seen by a doctor. When hospitalized, she visited him. When he needed a place to recover, she cared for him on the farm. While some may say that she only valued him for his ability to work the land, he believed otherwise. He appreciated her care and what he called her "warm heart." In later years, he would remind critics of the German people that he knew Germans who behaved kindly and well. Not everyone was a Nazi, nor was everyone hard-hearted. As Wasyl explained, "Not all Germans are bad people."

At war's end, Wasyl decided to leave the farm. Though he appreciated everything the farmer's wife had done for him, he wanted to be with his own people. The first step to get out of Germany was to report to a Displaced Persons (DP) camp. Landstuhl DP camp, packed with Polish refugees, was closest. Though not overly unfriendly, the administrators were understandably concerned first with the welfare of their own people.

The camp also teemed with Russian refugees, and soon Russian agents. In fulfillment of commitments made at the Yalta Conference, Allied authorities allowed Russian political and military officers access to the camps. The Russian officers were looking for Russian prisoners of war. The Russians considered their prisoners of war cowards who had betrayed the Motherland by not fighting to the death. Their fate was a murky one. They were also identifying anyone who might pose a threat to the Russian plan to dominate

Poland. The promise of free and open elections in Poland, part of the Yalta accords, convinced many Poles to return to their homeland. Promises of peace and prosperity under the flag of Polish communism rule convinced others. Unfortunately, Poland was under the Russian thumb. Promises of free and open elections were broken, much to the dismay of the British and Americans. Polish veterans were considered particularly worrisome; the fate of any identified Polish officers was the firing squad. Professional and educated persons were considered intelligentsia; many of the most capable and distinguished Polish ex-patriots were murdered.

The Russian agents were also looking for peasants capable of labor. Like Wasyl Soduk, taken from his village in the Ukraine in 1942 to work in Germany, the Russians wanted peasants to work the Soviet land, to produce the food for the Stalinist regime. Although Winston Churchill was not fooled by Stalin's promises, too many political and military leaders in both England and the United States took Stalin at his word. That is one of the main reasons Allied forces stopped at the Elbe in April 1945, leaving the capture of Berlin to the Russians. Within days of the end of the war, the Russians began to break promise after promise.

Wasyl knew better. He was not going to return to Russian control. He knew the realities of Bolshevism. His family had endured the reality of collectivization, the Stalinist policy of systemic starvation that killed millions in the early 1930s. At this time Wasyl also learned of the death of his brother, Mykola. When the Russians advanced into the Balkans in 1945, they were desperate to capture Hungary, Czechoslovakia, and Austria before the Allies. Proceeding with little regard for the human cost, Russian generals ordered waves of soldiers "over the top" into murderous machine-gun fire reminiscent of the assaults of World War One. In one such assault, Mykola was wounded near Budapest, Hungary. Rusia Soduk learned of her son's wounding and left for Budapest to try to find him and nurse him back to health, but to no avail. Mykola died in hospital, another soldier sacrificed for the Motherland.

Bad Kreuznach was in the French zone, the result of the Pots-dam Agreements of 1945. Spared destruction by an agreement between German authorities and the Commanding Officer of the advancing American forces under General Devers of the Sixth Army Group, Bad Kreuznach was not only a DP camp full of Ukrainians, it was also a German prisoner of war camp, or a Rheinwiesenlager. A beautiful city with a sad history, the "Field of Misery" memorial near the site of the camps reminds today's visitors of the sorrow experienced by so many.

Wasyl remained in Bad Kreuznach for almost five years. The French trusted him enough to make him part of the camp's Home Guard. With this status, he was issued a French army uniform and a Mauser rifle. However, the French would go only so far. The rifles were only for show; no bullets for the rifles were distributed. The French understood the depths of Ukrainian patriotism.

Despite vague hopes for the creation of an independent Ukrai-nian state, Wasyl knew it was unlikely. It was acceptable to honor the memory of Ukrainian patriots who declared an independent Ukrainian nation in October 1919. It was acceptable to lay flowers in honor of the Kiev patriots, but letters from home detailed the stran-glehold the ruthless Russians maintained on the village of Monas-teretch, forcing the farmers to give grain and livestock to the Bolshe-viks at every turn, without payment or compensation. He longed for his family but could not go back to his village near Lvov.

Wasyl blamed the Russians for Mykola's death. Rusia bore another son after the war and named him Mykola, in honor of her deceased son. His birth might have eased her sorrow, but it could never be the same.

Wasyl also knew World War Two was not really over. Stalin was determined to create a buffer zone between Mother Russia and the

West. This buffer zone would include Germany, Czechoslovakia, Hungary, and the Baltic States. The "greater good" would require continued sacrifices. Stalin was determined to conquer and occupy these countries. England, France, and the United States lacked the political will to resist his imperialist designs. Wasyl knew the Russians would use men in the future for mass attacks, with casualty rates that would never be tolerated in a democracy like England or the United States. Stalin used up men as the Allies used artillery shells, and Wasyl remained determined to avoid a return to Russian authority.

Meanwhile, the Russian authorities were agitating for the return of their men and women, especially those taken from Ukraine to Germany, but the Allies adopted a passive approach to the Russian demands; the Allies were not going to fight, but they were not going to cooperate, either, much to Russia's irritation.

Wasyl received offers to go to Canada, Australia, and South America from the International Red Cross. He declined them all. He did not want to go to any place but America. The American soldiers had treated him well, with respect. He applied with the American Red Cross for a sponsor. It would take years.

In the meantime, he met Anna, née Criczek, a Ukrainian from the village of Olesha. In the summer of 1942, Anna went to draw water from a local well, where she was abducted by a local German sympathizer. Since women were deemed more capable of the meticulous work needed to produce munitions, she was sent to an ammunition factory in Hamburg. For nearly three years she labored, producing bullets, bombs, and artillery shells. The living conditions were primitive; open barracks housed the women. Nightmares, screams of terror, and sobs of despair echoed throughout the facility night after night. Bombing raids by the British were frequent events that terrorized the women and pushed many to the depths of despair.

Resistance was futile. Should a worker fail to meet her quota, her feet were put into water and painful electrical shocks administered.

Attempts at escape resulted in public humiliation with the shaving of the escapee's head. For repeat offenders, especially those that might try and organize mass resistance, the punishment was cutting off of the breasts. Crude mastectomies failed to heal, became infected, and reminded all of the price to pay for resistance. For those women whose wounds healed, disfigurement compounded the realization that any baby they had would likely starve in infancy. For many women, the strain was too much. Suicide by hanging was the last resort for all too many.

Anna witnessed these atrocities and more, committed by both Germans and Russians. But like Wasyl, she also witnessed decent acts under difficult circumstances. When the German Army rolled through her village in 1941, occupation forces were left behind, including a German soldier named Hans, who was assigned to live in her home. Anna's family soon realized that this young man, filthy dirty and lice-covered, was scared and lonely. They treated him as well as they could. Fed, bathed, clothes laundered and disinfected, the soldier remained a part of the household, billeted under their roof, for nearly three years. He never laid a hand on Anna or any of her sisters.

Then one day in 1943, the soldier informed the family that he was leaving because the Russians were coming. He thanked the family for their many kindnesses, wished them the best of luck; with tears in his eyes, he departed, never to return. When the Russians rolled in, they were as ruthless to the German soldiers and the local Ukrainians as they had been in 1932. The small detachment of Germans was wiped out. The Russians believed the villagers had sympathized with their German occupiers. If not, why hadn't the villagers resisted? Because the villagers had seemingly collaborated with the Germans, the Russian soldiers were merciless as they searched for food, ransacking homes and barns, stealing everything of value they could find, including household goods and farm animals. Desperate efforts were made to hide animals from the

Russians. Anna's father went so far as to dig a trench in the woods behind their farm in which to hide a cow. Anna's sister Sofia tended the cow day and night. During the day, the trench was covered over with tree branches and sod. At night Sofia took the cow out of the trench and into the pasture to graze. Never letting the cow out of her sight, she remained "on guard" for Russian soldiers foraging in the village. She milked the cow twice a day. The milk made the difference between life and death for her family. When the Russian troops moved west in pursuit of the retreating Germans, she was able to return home with her cow to her grateful family.

Even simple, decent actions were viewed by the Russians with suspicion. After the Russians left, the villagers buried the bodies of German soldiers next to the local church, they hid that fact from the Russians out of fear the Russians would disapprove and exact reprisals. The Russians were ruthless and showed the villagers no pity.

So, in 1945, when the British finally liberated Hamburg, and despite the difficulties involved, Anna wanted to go to America. When she finally met Wasyl in 1946, Anna's determination was confirmed. Once they met, they never separated. They held the same values, the same dream of America. In 1951 a Ukrainian church in Canton, Ohio, through the efforts of Katherine Skubik and her son Stephen Skubik, a U.S. Army counterintelligence agent, agreed to sponsor the young family. From the camp, Anna and Wasyl traveled to Bremen, Germany for passage to New York. From there, they had a two-day train ride to Canton. The church had a house for them to live in, a block away from Wean United, a steel company that needed men capable of working in a steel mill, a dangerous world of molten steel, heat, and dust. Wasyl reported for work on the following Monday. Within a month, to consecrate their union, and to honor the church, they were married in St. Nicholas Ukrainian Catholic Church in Canton, Ohio.

Nine months later, son Walter "Walt" was born. Walt was Wasyl and Anna's second child, born in the United States. Daughter Marie was born in 1950 in the displaced persons' camp in Germany. Anna was illiterate and would remain so all her life despite the repeated efforts of her daughters, Marie and Olga, teachers by training. Wasyl and Anna educated their three children, all of whom graduated from college and professional schools, raising families of their own.

In 1988, shortly after the fall of the former Soviet Union, Walt and his wife, Deborah, traveled to his father's village of Monasteretch. He wanted to see his father's hometown for himself and to visit his father's sisters who still lived in the village. When he was there, an older gentleman came up and said that he had been the mayor in 1942. He wanted to explain. At first, he said, villagers volunteered to go to Germany to help with the war effort. But after a few months, there were no more volunteers. Instead, the mayor was expected to choose people to fill the quota of able-bodied men the Germans required every month. Since it was able-bodied men they wanted, he fully expected that the Germans would reject the small and scrawny Wasyl Soduk. He was wrong, and for that he was sorry and wanted to apologize.

Anna and Wasyl would wait more than 50 years before returning home to Ukraine for a visit in 1992. Some of Wasyl's family, nieces and nephews, were still living in the village of Monasteretch. Wasyl wanted to see his old home, his old friends, and family. So it was arranged. For the villagers, it was a big deal. Very few visitors ever came from the West, and never from the United States. People had heard stories of the long-lost Wasyl Soduk, but were they really true? While Wasyl wandered the streets, people came out of shops and stores to see if it was really him. Of course, with the first words out of his mouth, folks knew he had made it back.

I had the privilege of meeting Wasyl soon after his Ukraine trip, around 1990. He was feeling weak and run down. At first, it was ascribed to a life of hard work and worry. A man of nearly 70 would naturally slow down, even one as vigorous as Wasyl, right? Well, I put on my medical hat and began my investigations. A blood test showed he was anemic and most likely iron deficient. Iron-deficiency anemia is often a sign of something bad, especially in a 70-year-old man. So I proceeded with an upper endoscopy, an examination of the stomach and intestines, looking for lesions like ulcers that can cause chronic gastrointestinal blood loss. What I discovered was a Billroth II reconstruction. Turns out, a German surgeon operated on Wasyl in January 1948, performing a Billroth II reconstruction.

Every general surgeon today knows the name of the German surgeon Dr. Billroth, who in the 1930s pioneered the ulcer operation named after him. One consequence of the Billroth II surgery is an inability to absorb iron efficiently from one's diet. So the likely explanation for Wasyl's anemia was iron-deficiency secondary to chronic malabsorption of iron as a consequence of his previous surgery.

How ironic it seemed to me. Wasyl, a Ukrainian taken by the Germans as a slave laborer, underwent a life-saving surgery in 1948 for what was likely chronic ulcer disease (pain, intermittent gastrointestinal bleeding), the surgery performed in a former German military hospital.

The last part of my evaluation included a colonoscopy to exclude any other pathology of the gastrointestinal tract that might be contributing to Wasyl's anemia. Turns out, at colonoscopy I found a right-sided colon cancer. Another surgery, an easy recovery, and Wasyl's long life continued.

I have had the privilege to know not only Wasyl but his family. On a second trip to the Ukraine in the mid-1990s, father and son journeyed to Anna's village, Olesha. Alas, it was gone, burned to the ground by the Russians in 1945. The villagers of Olesha pointed out the mounds of earthworks, the machine gun nests, constructed by the German soldiers in their vain attempt to hold off the Russian attack. Walt also noticed a new stucco church in the village near the wooden church more characteristic of the region. He asked, "When did you build a new church?"

The reply: "We built the new church in 2005, with money from the German government. When the German Overseas Graves Commission learned that we had buried the 15 or so German soldiers in 1944 next to our little wooden church, they sought to repatriate the bodies for burial in Germany. In a gesture of appreciation and reconciliation, they paid for the construction of this new church."

It made sense. The fancy, heated church for the village of Olesha was accepted as atonement for the occupation of the village so many years before. Such an important gesture, no less important than the explanation and apology offered to Wasyl by the mayor. The mayor had made a judgment that for the village, it was better to lose a scrawny kid, not someone more likely to survive. Ironically, the work in the woods, chopping trees, dodging American airplane bullets, and surgery by a German military surgeon saved Wasyl's life. The German government likewise made amends to the soldiers it sent to Russia, and to the soldiers' families.

Long into his 90s, Wasyl Soduk was tending his garden and still telling his great-grandchildren stories about life in the old country, about his journey to the United States, and about the value of freedom. I count myself fortunate to have met him and to have helped him continue to live a long and productive life. I am also grateful to the Soduk family for sharing the intimate details of their family's story, a story of courage, fortitude, and perseverance.

EPILOGUE

My grandparents' home on Albemarle Street in Westmoreland Hills of Washington, D.C., was a repository of history, some of it only now being told. Letters, photographs, and artifacts, on the front hall table, on the foyer wall, or placed in a trunk in the attic, were all waiting to tell a story. From the time I was a six-year-old until the emptying of the house for sale 30 years later, the treasure chest of family life, the ups and downs, the achievements and the disappointments, and the loves and the hates were present, waiting to be explored and understood.

The picture of General Robert E. Lee in the front foyer, the daguerreotype on the side table, or the silver tray in a chest of drawers were but tips of icebergs. On a side table in the living room sat a fancy enameled medal with crossed swords on a bright orange ribbon in a formal presentation box. As a child, I thought it pretty. I'd pick it up, surreptitiously placing it around my neck, for even then I knew the medal was special and not meant for little boys and their careless ways.

Only later, while writing about my grandfather in my first book, *General in Command: The Life of Major General John B. Anderson; from Iowa Farm to Command of the Largest Combat Corps in World War II*, did I come to know that the Order of the Orange Nassau, Grand Officer, with Swords was the second-highest military award given by the Netherlands. I knew nothing of the significance of Netherlands' Queen Beatrice's expression of gratitude until I met a Dutchman in the desert in 2015, and then Janice Amos, the daughter of Sgt. Von Henke, in Roermond, in 2020. Learning about the magnitude of this honor bestowed upon my grandfather was yet another gift he gave me, a gift of pride in his legacy.

Finishing unfinished business has motivated me all along in my writing. Early in the process, others asked what I hoped to accomplish. "I wanted to know my grandfather as a man, not just my Granddaddy," I would reply. I would sometimes add, "I want my grandfather's legacy of service to endure. I want him to be placed among the pantheon of combat leaders of World War Two." An exhibit devoted to Anderson's life was created for the MAPS Air Museum in Canton, Ohio, a chair dedicated to him is in the U.S. Army Museum at Fort Belvoir, and his Cullum file has been written. All appropriate and gratifying accomplishments, but there is more.

A corps commander is usually a three-star general. During the war, many of the Army corps commanders wore only two stars, but most received a third star at the end of the war, in recognition of their service. Granddaddy never received a third star, not at the end of the war, and not upon retirement. The other two corps commanders in Simpson's Ninth Army, Generals Gillem and McLain, were promoted to three-star lieutenant general at the end of World War Two. General Simpson wrote to my grandfather "Andy" in 1958:

> Dear Andy:
>
> Your letter of June 30 was received and I was delighted to hear from you again.
>
> A few years ago, I wrote you that during the war I had recommended your promotion to lieutenant general. I would like to repeat what I told you then.
>
> During World War Two you commanded the XVI Corps which was one of the three Army Corps permanently assigned to the Ninth Army which was commanded by me. You and the XVI Corps under your command performed in a superior manner during all combat operations which included an attack from the Roer River to the Rhine River, the crossing of the Rhine River, the capture of the Ruhr area, and the encirclement and capture in cooperation with units of the First Army of a pocket of about 350,000 German soldiers.
>
> Because of your splendid record and performance of duty during the entire time you were under my command, I

recommended you for promotion to the grade of lieutenant general and I was informed that this recommendation was approved by General Bradley who commanded the XII Army Group of which the Ninth Army was a part.

I have always regretted that you did not receive this promotion which you so richly deserved. I wish to assure you that it was an honor and a pleasure to serve with you and your fine XVI Corps, and I shall always feel most grateful to you for the fine job that you did....

Sincerely,
Bill
William H. Simpson, General,
U.S. Army, Retired

I tried to secure a posthumous promotion for Granddaddy. Since Granddaddy was an Iowa native, I thought congressional support from Iowa might be helpful. Mr. Rob Maharry of the Parkersburg, Iowa *Eclipse News Review* informed me that Iowa Senator Charles Grassley's home was only 10 miles from Granddaddy's hometown of Parkersburg. When approached, Senator Grassley's office was supportive of the effort; in fact, Senator Grassley was kind enough to write the cover letter asking for Granddaddy's promotion. Legislative aide Mrs. Liesel Crocker guided me through the process of record retrieval and compilation. I wrote a detailed summary of Granddaddy's career.

The request was denied by the Army Board of Corrections with a letter signed by a brigadier general with the terse explanation that promotion is given "not for service performed but for future assignments."

I was disappointed, but I take comfort in having made the effort, and in the words of Teddy Roosevelt that I remember Granddaddy reading me years ago:

> It is not the critic who counts; not the man who points out how the strong man stumbles, or where the doer of deeds could have done them better. The credit belongs to the man who is actually in the arena, whose face is marred by dust and

sweat and blood; who strives valiantly; who errs, who comes up short again and again, because there is no effort without error, and shortcoming; but who does actually strive to do the deeds; who knows great enthusiasms, great devotions; who spends himself in a worthy cause; who at the best knows in the end the triumph of high achievement, and who at the worst, if he fails, at least fails while daring greatly, so that his place shall never be with those cold and timid souls who neither know victory nor defeat.

As mentioned earlier, on the occasion of the 75th anniversary of the liberation of Roermond, I had a chance to say a few words to the large crowd gathered for the occasion in the town cathedral. It is a beautiful building, rebuilt by the citizens of Roermond after the war, the spire literally rising from the ashes. I spoke of the brave men and women of the American Armed Forces, especially the men of the 15th Cavalry who first entered the city. If Granddaddy had been there, I believe he would have praised the strength and endurance of the Dutch who resisted Nazi occupation for five years and emerged scarred but whole, determined to get on with their lives, to pursue their loves and dreams, and to never forget the horrors of tyranny.

These are the reasons I am inspired to continue to tell stories of heroes—those from my family and others. When I ask friends or acquaintances, "What did your parents and grandparents do in World War Two?" stories pour out. I encourage you to ask that question, as well, and to listen, record, and share the stories you hear, too. I believe it is our collective responsibility to write the next chapter in the cause of freedom. The editors of *The American Yawp* acknowledge that we are the heirs of our history but add "every generation must write its own...." Maybe so. If so, the responsibility is a large one. Whereas conclusions about slavery and the motives of European explorers bear re-examination, the self-evident truths at the heart of Western philosophy are more difficult things to throw in the trash bin of history.

I recently came across a speech by former President Theodore Roosevelt given in November 1918 that paradoxically gives me hope and yet leaves me discouraged. Hope, because the message is timeless, and discouraged, because the message is now more than 100 years old. Here's what he said:

> Now, friends, both the white man and the black man in moments of exultation are apt to think that the millennium is pretty near; that the sweet chariot has swung so low that everybody can get upon it. I don't think that my colored fellow citizens are a bit worse than my white fellow citizens as regards that particular aspiration. And I am sure you do not envy me the ungrateful task of warning both that they must not expect too much. They must have their eyes on the stars but their feet on the ground. I have to warn my white fellow citizens about that when they say: "Well, now, at the end of this war we are going to have universal peace. Everybody loves everybody else." I want you to remember that the strongest exponents of international love in public life today are Lenin and Trotsky.
>
> To each side I preach the doctrine of thinking more of his duties than of his rights. I don't mean that you shan't think of your rights. I want you to do it. But it is awfully easy, if you begin to dwell all the time on your rights, to find that you suffer from an in-growing sense of your own perfections and wrongs and that you forget what you owe to anyone else.
>
> I congratulate all colored men and women and all their white fellow Americans upon the gallantry and efficiency with which the colored men have behaved at the front, and the efficiency and wish to render service which have been shown by both the colored men and the colored women behind them in this country.

Roosevelt preached a hard lesson. So did our fathers and grandfathers. We would do well to contemplate their stories lest we make the same mistakes that led to two world wars and a century of political and racial strife.

I hope *To the Front* will remind you in a small way of the enduring words of Abraham Lincoln:

With malice toward none, with charity for all, with firmness in the right as God gives us to see the right; let us strive on to finish the work we are in; to bind up the nation's wounds; to care for him who shall have borne the battle, and for his widow and his orphan—to do all which may achieve and cherish a just and lasting peace, among ourselves, and with all nations…

And the words of Winston Churchill:

This is the lesson: never give in, never give in, never, never, never, never—in nothing, great or small, large or petty— never give in except to convictions of honor and good sense. Never yield to force; never yield to the apparently overwhelming might of the enemy…

And John F. Kennedy's inspiring words:

Ask not what your country can do for you—ask what you can do for your country.

And the words of Dr. Martin Luther King, Jr:

I have a dream that my four little children will one day live in a nation where they will not be judged by the color of their skin, but by the content of their character.

Lastly, it is time for all Americans to recall and embrace the legacy that the greatest generation bequeathed us: a free nation protected with a terrible sacrifice in blood and unknowable personal torment, and a pride that will never accept surrender or submission. We are a free and proud people who triumphed during World War II and whom President Kennedy summoned in 1962. Our war against subversion presents challenges that America has never faced. The enemy frequently is elusive, and the rules of engagement are not well-defined. But the bugle has sounded, calling us to find our way forward and to the battle join.

* * *

WORKS CITED AND CONSULTED

Abelow, Samson Z. *History of the XVI Corps: From Activation to the End of the War in Europe.* Washington, D.C., Infantry Journal Press, 1947.

Albert, Susan Wittig. *The General's Woman: A Novel.* Bertram, TX, Persevero Press, 2017.

Ambrose, Stephen E. *D-Day June 6, 1944: The Climactic Battle of World War II.* New York, NY: Simon and Schuster, 1994.

Arnn, Larry P. *Churchill's Trial: Winston Churchill and the Salvation of Free Government.* Nashville, TN, Nelson Books, 2015.

Baldridge, Robert C. *Victory Road.* Bennington, VT: Merriam Press, 1999.

Bolton, Martha and Hope, Linda. *Dear Bob: Bob Hope's Wartime Correspondence with the G.I.s of World War II.* Jackson, MS, University Press of Mississippi, 2021.

Bradley, James. *The Imperial Cruise.* New York, NY, Little, Brown and Company, 2009.

Bradley, Omar. *A Soldier's Story.* New York, NY, Henry Holt, 1951.

Butcher, Harry C. *My Three Years with Eisenhower.* New York, NY, Simon and Schuster, 1946.

Chandler, Alfred D. *The Papers of Dwight David Eisenhower.* Baltimore, MD, Johns Hopkins Press, 1970.

Cirillo, Roger. *The Campaigns of World War II: Ardennes-Alsace.* U.S. Army Center for Military History, Lightning Source, UK, 2011.

Clarke, Bruce C., *The Battle at St. Vith, Belgium, 17–23 December 1944.* The U.S. Army Armor School, 1966.

Coffman, Edward M. *The War to End All Wars.* Madison, WI, University of Wisconsin Press, 1986.

Cohen, Roger. *Soldiers and Slaves.* New York, NY, Knopf, 2005.

De Varila, Corporal Osborne. *The First Shot for Liberty.* Philadelphia, PA, John C. Winston Co, 1918.

Deisseroth, Karl. *Projections.* New York, NY, Random House, 2021.

Dupuy, R. Ernest. *St. Vith: Lion in the Way, The 106th Infantry Division in World War II*. Nashville, TN, The Battery Press, 1986.

Eisenhower, John S.D. *Yanks*. New York, NY, The Free Press, 2001.

Enright, Dominique. *The Wicked Wit of Winston Churchill*. London, Michael O'Mara Books, 2001.

Felton, Mark. *Ghost Riders: When U.S. and German Soldiers Fought Together to Save the World's Most Beautiful Horses in the Last Days of World War II*. New York, NY, Da Capo Press, 2018.

Fenelon, James M. *Four Hours of Fury*. New York, NY, Scribner, 2019.

Ferrell, Robert H. *The Eisenhower Diaries*. New York, NY, W.W. Norton, 1981.

Forty, Simon. *Across the Rhine: January–May 1945*. Havertown, PA, Casemate, 2020.

Frey, Richard L. *According to Hoyle: Official Rules of More than 200 Popular Games of Skill and Chance with Expert Advice on Winning Play*. New York, NY, Fawcett Books, 1956.

Fye, John Harvey. *History of the Sixth Field Artillery 1798–1932*. Harrisburg, PA, Telegraph Press, 1933.

Gerard, James L. *My Four Years in Germany*. New York, NY, Grosset & Dunlap, 1917.

Harries, Meirion and Susie. *The Last Days of Innocence*. New York, NY, Random House, 1997.

Helprin, Mark. *A Soldier of the Great War*. Orlando, Florida, Harcourt Brace Jovanovich, 1991.

Holland, James. *Brothers in Arms: One Legendary Tank Regiment's Bloody War from D-Day to VE-Day*. London, UK, Penguin Random House, 2021.

Johnson, Ian Ona. *Faustian Bargain*. Oxford, UK, Oxford Press, 2021.

Kershaw, Alex. *The Longest Winter: The Battle of the Bulge and the Epic Story of World War II's Most Decorated Platoon*. Cambridge, MA, Da Capo Press, 2004.

Kershaw, Robert. *Landing on the Edge of Eternity: Twenty-four Hours of Omaha Beach*. New York, NY, Pegasus Books, 2018.

Kilborne, Al. *Woodley and Its Residents.* Charleston, SC, Arcadia Publishing, 2008.

King, Martin, Johnson, Ken and Collins, Michael. *Warriors of the 106th: The Last Infantry Division of World War II.* Philadelphia, PA, Casemate, 2017.

Larson, Erik. *Dead Wake.* New York, NY, Crown Press, 2015.

Lavin, Frank. *Home Front to Battlefront: An Ohio Teenager in World War II.* Athens, OH, Ohio University Press, 2016.

Lewis, C.S. *Surprised by Joy: The Shape of My Early Life.* London, UK, Harvest/HBJ, 1956.

Locke, Joseph L. and Wright, Ben. *The American Yawp: A Massively Collaborative Open U.S. History Textbook.* Stanford, CA: Stanford University Press, 2019.

Long, Jeff. *Dual of Eagles.* New York, NY, William Morrow and Company, 1990.

MacDonald, Charles. *A Time for Trumpets: The Untold Story of the Battle of the Bulge.* New York, New York, William Morrow and Company, 1985.

Mansoor, Peter R. *The GI Offensive in Europe: The Triumph of American Infantry Divisions, 1941–1945.* Lawrence, KS, University Press of Kansas, 1999.

Margetts, Colonel *Nelson E. A History of the Sixth Regiment Field Artillery First Division United States Army.* Ransbach, Germany, 1919.

Martianoff, N. and Stern, M.A. *Almanac of Russian Artists in America, Volume One.* New York, NY: Martianoff and Stern, 1932, p. 89.

McCullough, David. *The Pioneers: The Heroic Story of the Settlers Who Brought the American Ideal West.* New York, NY: Simon and Schuster, 2019.

McManus, John C. *The Dead and Those About to Die: D-Day: The Big Red One at Omaha Beach.* New York, NY, Dutton Caliber, Penguin Random House, 2014.

McManus, John C. *Alamo in the Ardennes: The Untold Story of the American Soldiers Who Made the Defense of Bastogne Possible.* New York, NY, Dutton Caliber, Penguin Random House, 2007.

Merriam, Robert E. *Dark December: The Full Account of the Battle of the Bulge.* First published 1947, Los Angeles, CA, Enhanced Media Publishing, 2017.

Mick, Allan H. *With the 102d Infantry Division Through Germany.* Washington, D.C., Infantry Journal Press, 1947.

Millard, Candice. *The River of Doubt.* New York, NY, Doubleday, 2005.

Moorehead, Alan. *Eclipse.* New York, NY, Harper & Row, 1945.

Morelock, Jerry D. *Generals of the Bulge: Leadership in the U.S. Army's Greatest Battle.* Mechanicsburg, PA, Stackpole Books, 2015.

Morgan, Kay Summersby. *Past Forgetting: My Love Affair with Dwight D. Eisenhower.* New York, NY, Simon and Schuster, 1975.

Neumann, Ariana. *When Time Stopped: A Memoir of My Father's War and What Remains.* New York, NY: Scribner, 2020.

Nicole, John and Rennell, Tony. *The Last Escape: The Untold Story of Allied Prisoners of War in Europe 1944–45.* New York, NY, Viking, 2003.

Oldfield, Barney. *Never A Shot Fired in Anger.* New York, NY, Duell, Sloan and Pearce, 1956.

Pergrin, David. *Engineering the Victory: The Battle of the Bulge.* Atglen, PA, Schiffer Military/Aviation History, 1996.

Prange, Gordon W. *Miracle at Midway.* Norwalk, CT, Easton Press with permission of McGraw-Hill, 1982.

Prefer, Nathan N. *The Conquering Ninth: The Ninth U.S. Army in World War II.* Philadelphia, PA: Casemate, 2020.

Quarrie, Bruce. *The Ardennes Offensive: V U.S. Corps and XVIII U.S. (Airborne) Corps.* Oxford, U.K., Osprey, 1999.

Radford, Albert E. and Radford, L.S. *Unbroken Line: The 51st Engineer Combat Battalion—From Normandy to Munich.* Woodside, CA, Cross Mountain, 2002.

Ramsey, Raquel and Aurand, Tricia. *Taking Flight: The Nadine Ramsey Story.* Lawrence, Kansas, University of Kansas Press, 2020.

Reichert, Walter E. *Phantom Nine: The 9th Armored (Remagen) Division 1942–1945.* Presidial Press, 1987.

Rixey, Randolph Picton. *The Rixey Genealogy.* Lynchburg, VA, JP Bell, 1933.

Roscoe, Theodore. *United States Destroyer Operations in World War II.* Annapolis, Maryland, United States Naval Institute, 1953.

Ryan, Cornelius. *The Last Battle.* New York, NY, Simon and Schuster, 1966.

Scott, Emmett J. *Scott's Official History of the American Negro in the World War.* Homewood Press, 1919.

Snyder, Timothy. *Bloodlands: Europe Between Hitler and Stalin.* New York, NY, Basic Books, 2010.

Stelpflug, Peggy A. and Hyatt, Richard. *Home of the Infantry: The History of Fort Benning.* Macon, GA, Mercer University Press, 2007.

Townsend, Tim. *Mission at Nuremberg.* New York, NY, Harper Collins, 2014.

Truscott, Lucian K. *The Twilight of the U.S. Cavalry.* Lawrence, KS, University Press of Kansas, 1989.

Tuchman, Barbara W. *The Zimmermann Telegram.* New York, NY, Macmillan, 1958.

Tyng, Charles. *Before the Wind: The Memoir of an American Sea Captain 1808–1833.* New York, NY, Viking, 1999.

Tzouliadis, Tim. *The Forsaken: An American Tragedy in Stalin's Russia.* New York, NY, Penguin Press, 2008.

United States Army. *The 2nd Infantry Division.* Nashville, TN, Battery Press, 1979.

----- *Conquer: The Story of the Ninth Army, 1944-1945.* Washington D.C., Infantry Journal Press, 1947.

----- *The 102nd Infantry Division.* Atlanta, GA, Albert Love Enterprises, 1947.

----- *The Cross of Lorraine: A Combat History of the 79th Infantry Division, June 1942–December 1945.* Unknown publisher, 1946.

Van Ness, Michael M. *General in Command: The Life of Major General John B. Anderson; from Iowa Farm to   Command of the Largest Combat Corps in World War II.* Virginia Beach, VA, Koehler Books, 2019.

Varila, Corporal Osborne de. *The First Shot for Liberty: The Story of an American Who Went Over with the First Expeditionary Force and Served His Country at the Front.* New York, NY: Grosset & Dunlap, 1918.

Waugh, Evelyn. *The Sword of Honor Trilogy: Men at Arms*. New York, NY, Everyman's Library by Alfred A. Knopf, 1994. P.36. (Note: *Men at Arms* first published 1952.)

Waugh, John G. *The Class of 1846*. New York, NY, Warner Books, 1994.

Wawro, Geoffrey. *Sons of Freedom: The Forgotten American Soldiers Who Defeated Germany in World War I*. New York, NY, Basic Books, 2018.

Willoughby, Lynn. *Judge Aaron Cohn: Memoirs of a First Generation American*. Lexington, KY, self-published, 2008.

Wilmot, Chester. *The Struggle for Europe*. London, UK, Collins, 1952.

Wilson, Clyde. *The Landon School Story*, 1968.

Wood, Kieron. *Ike's Irish Lover: The Echo of a Sigh*. Lexington, KY, self-published, 2019.

Werth, Nicolas, et al. *The Black Book of Communism: Crimes, Terror, Repression*. Cambridge, MA, Harvard University Press, 1999.

INDEX

Page numbers in *italics* refer to photographs or illustrations.

B-25 Mitchell bombers, 22
B-29 bombers, 207
B-52 bombers, 207
Baker, Newton (Secretary of War), 77–78
Band of Brothers (television series), 159, 160, 161, 168
Banfield, Mary Lee, *128*
Banfield, Paul Landon, 16, 18, 20–21, 58, 65–66, 68, 111, *128*
Barron, Leo, 168
Bastogne, siege of, 60–61, 152, 153–169
   map of, 162
Bataan Death March, 24, 26
Bates, Mr., 61
Bayerlein, Fritz (German General), 157
Beatrice (Dutch Queen), 255
Beauvoir School, 11–13
Bell, Samuel, xix
Bennett, James B. (Chief Warrant Officer), 141–142
Bennett, Paul G. (Private), 135
Berga labor camp, 143–145
Berga: Soldiers of Another War (film), 143
Berle, Adolf, 72
Berlin Airlift, 187
Bethesda Naval Hospital, xx, xxi, 47, 49, 106, *129,* 197, 210, 215, 217–
   218, 229–230
Billington, R.C. (Sergeant), 149
Billroth II surgery, 253
Bismarck (German battleship), 204
Black Book of Communism, The (book), 240
Bloodlands (book), 237
Boggess, Charles P. (Lieutenant), 169
Bolshevism, xiii, 87, 97, 132, 236–241, 247–248; *see also* Soviet Union
Bolshoi Ballet, 194
Bradley, Omar (General), 15, 92, 137, 153–154, 158, 166, 173–174, 185,
   186, 187, 194, 195, 210, 257
   Twelfth Army and, 173, 185
Brandeis University, 16
Brigade of Midshipmen, 204
British military
   21st Army Group, 173, 185
   Royal Air Force (RAF), 22, 124, 168, 243
   XII Corps, 137, 175
   XXII Corps, 51
Bronze Star, 143
Brooke, Alan, *see* Alanbrooke, Lord (British Field Marshal)
Brothers in Arms (book), 184

Brown, Lloyd (General), 155
Brown v. Board of Education (Supreme Court case), 15
Buchanan, James (President), 72
Bulge, Battle of the, 21, 60, 94, 102, 119, 148, 153, 170, 171–178, 179,
    192–193
Bullard, Robert Lee (Lieutenant General), 55
Bullitt, William C., 88
Bundy, McGeorge (United States National Security Advisor), 98
Burke, Edmund, 222
Busby, Matt, 66
Bush, George H.W. (President), 175, 233
Bush, George W. (President), 175, 188
Bushido Code, 24
Byron, Lord, 74, 89

C
Cabanatuan Camp #1 (Philippines), 24–25
"Caisson Song, The," 170
Camp Douglas (Illinois), 81, 85–86, 112
Camp Dust (Texas), 34
Camp Perry (Ohio), 77
Camp Wachusett (New Hampshire), 170
Cantigny, Battle of, 55–56
Canton, Ohio, 32, 235, 251, 256
Canton Bulldogs, 32
Carlisle Indian School, 33
Carlson, Arthur A. (Private First Class), 108
Carter, Jimmy (President), 212
Castell, Donald, 106
Catherine II (Russian Empress), 239
Catholicism, 19, 89, 208
CBS Evening News, 171–172
Chapin, Fred W. (Corporal), 149, 151
Chevy Chase Country Club, 11, 98
China, 33, 188, 201
Chobanian, John S., 233–234
Chobanian, Sarkis J., 230, 233–234
Churchill, Winston S. (British Prime Minister), 6, 64–65, 91–92, 94, 97,
    121, 172–173, 183, 186–189, 203, 247, 260
  Rhine River crossing and, 103, 186–189
Churchill War Cabinet Rooms, 188–189
Civil Rights Act (1964), 18
Civil Rights Movement, 16, 18
Civil War, xix, 16, 30, 70, 77–85, 88, 89, 112, 149, 170
Civil War (book), 84

Holmberg, Paul A. "Lefty" (Rear Admiral), *130,* 229–233
  gravesite of, 233
Holocaust, 43, 97, 244–245
Holodomor, 237–239
Hood, John Bell (General), 78
Hope, Bob, 195
Hopkins, Armand (Colonel), xxv, 22–27, 61, *128*
  captivity of, 24–26
  memoir of, 23
Houghton, Bill, 34, 36
Howitzer (West Point yearbook), 32
"Hukilau Song, The," 9–10
Hussein, Saddam, 175
Hustead, Charles L. (Major), 160

I
International Red Cross, 73–74, 249; *see also* Red Cross
Irwin, S. Leroy (Major General), 167
"Italian Holliday" (sketch), 90, *112*

J
Jackson, Andrew (General, President), 78
Jackson, Stonewall (General), 78
Jacques, George L. (Lieutenant Colonel), 169
James, Henry, 74
Jim Crow laws, 89
Johnson, Ian Ona, 236
Johnson, Lyndon B. (President), 188, 206
Johnson, Margaret Anderson, 56–57, 101
Jones, Alan W. (Major General), 176
Jones, Alvin (Major), 164–165
Jones, Gwen, 22
June, Carl, 47

K
Kalahari Desert, 189
Kaye, Milton, 194
Keating, Frank A. (General), 35, 177
Kellogg, Hamilton H. (Chaplain), 99, 102–103, 104
Kennedy, Ethel, 98
Kennedy, Jackie, 104, 107
Kennedy, John F. (President), xiii–xiv, 33, 104–108, *127,* 206, 260
  assassination and funeral of, 105–108
  inaugural address of, 104
Key, Philip Barton, 72

Manfred (book), 89
Maret School, 72
"Marseillaise, The," 42
Marsh, Richard Symmes Thomas (Lieutenant), 21, *120*
Marshall, George (General), 70, 93, 94, 95, 134, 137, 172
Mason-Dixon line, 78
Mauldin, William, 135
May, John, 22, *124*
Mayo, John H.F. (Private), 20
McArthur, Douglas (General), 72
McAuliffe, Anthony C. (General), 59–61, 65, 88, 157–168; *see also*
    Bastogne, siege of
McCullough, David, xx
McLain, Raymond S. (Major General), 153, 256
McLendon, Captain, 44
McMahon, Leo T. (Brigadier General), 138–139
McManus, John, 156
McNair, W.S. (Colonel), 45
Megee, Betsy, 102
Men at Arms (book), 195
Merriam, Robert, 151–152
Mexican–American War, xix, 79
Mexican Punitive Expedition, 34, 35, 37–38, 133, 136
Middleton, Troy (Major General), 135–136, 153–154, 156–160, 167
Midway, Battle of, 25, 130, 230–231
    map of, 232
Mikado, The (opera), 101
Miley, William "Bud" (General), 184
Military Aircraft Preservation Society (MAPS) Air Museum, *131,* 256
Milliken, Charles M., 33
Millikin, John (General), 167
Miracle at Midway (book), 230, 233
Model A Ford, 98–99
Molotov–Ribbentrop pact, 238
Montgomery, Bernard L. (British Field Marshal), 6, 92, 95, *121,* 132,
    172–175, 183–186
    21st Army Group, 173, 185
    command of American troops by, 172–175
Moore, Ned (Lieutenant Colonel), 164
Morgenthau, Robert (Lieutenant Commander), 93, *129,* 205–206
Mount Vernon Academy for Girls, 11
Mountbatten, Lord, 204
Mrs. Shippen's Dancing School, 98

Pentagon, xix, 70
Pergrin, David E. (Lieutenant Colonel), 146–147, 151
Perry, Oliver Hazard (Captain), 204
Perry, Tom, 59, 61
Pershing, John J. "Black Jack" (General), 34, 36, 38, 43, 54, 55, 72, 75, 173
    speech given to 1st Infantry Division by, 55
Pervitin (drug), 148
Philippines, 24–25, 31, 33, 70, 71, 77, 86, 133
Phillips, Captain, 222
Phillips Academy, 71
Pigg, Edwin (Staff Sergeant), 149
Pioneers, The (book), xx
Plattsburg Movement, 71; *see also* Preparedness Movement
Point Mugu, California, 9, 12
Poland, 90, 96, 143, 179, 236, 238–241, 244, 246–247
Porter, Ray E. (Major General), 177
Potemkin, Grigori, 239
Potemkin village, 239
Potsdam Agreements of 1945, 248
Powers, Bernie, 230
Prange, Gordon W., 230
Premetz, Ernest (Private First Class), 163, 165
Preparedness Movement, 70–71; *see also* Plattsburg Movement
Prickett, Fay B. (Major General), 177
Prior, John T. (Major), 163, 169
Prothro, Nancy, 9

Q
Queen Mary (ship), 64, 203
Quesada, Elwood (General), 88
Quo Vadis apartments, 109

R
Raaen, John (Captain), 155
racial segregation, 15–18, 89, 259–260
Radnor Elementary School, 15, 58
Ramcke, Hermann (German Generalleutnant), 135–136
Rapallo, Treaty of, 236
Reconstruction, 16, 85, 88–89
Red Army Chorus, 194
Red Cross, 24, 73–74, 143, 249
Regulus missile system, 9
Reminiscences of Armand Hopkins (memoir), 23
Reston, James, 238
Revolutionary War, *see* American Revolution

von Manteuffel, Hasso (General), 152, 154, 166
Vonnegut, Kurt, 143

W
Wacht am Rhein offensive, 148, 175
Wadden, Tommy, 61, 62, 63, 66–67
Wadsworth, Eliot, 73
Wagner, Major, 163, 165–166
Wainwright, Jonathan (General), 24
War of 1812, 78, 204
Washington, George (General, President), xv, 16, 98
Washington National Cathedral, 11, 72, 110
Washington Post (newspaper), 18, 86
Watson, Leroy H., 136–137
Waugh, Evelyn, 195
Weber, Richard E. (Lieutenant Colonel), 141–142
Webster, Duane J. (Second Lieutenant), 169
Wehle, Philip C. (Major General), 106
Weissberg, Alexander, 238
Werth, Nicolas, 240
West Point, *see* U.S. Military Academy
West Point Club of Washington, 59, 61
Wharton, James E. (Brigadier General), 155
When Time Stopped (book), 111
White Star Lines, 73
Williams, Hermann W., Jr., 84–85, 86
Williams, Samuel King (Major), 85–86
Willie and Joe (cartoon), 135
Wilson, Woodrow (President), 38, 44, 78
Wiltgen, Robert G. Lt. (jg), vi, 208
Wiltgen, Rosemary, 208
Wingo, Grace Amoleyetto, 36–39, 41, 69, *113*
women's suffrage, 68–69, 75
Wood, Leonard (General), 70–71
Woodley Mansion, 72–73
World War One, xix, xxi–xxii, 35, 42–45, 49–50, 71–72, 88, 133, 172, 173, 236, 240, 247; *see also* Versailles, Treaty of; *individual battles and officers*
    Christmas truce during, 168
    first shot fired during, 44–45
    Hundred Days Offensive of, 133
    start of, 37–39
World War Two, 14–15, 33, 61, 101, 104–105, 122, 132–191, 212, 233–234, 240–253, 256–257; *see also* Pearl Harbor, attack on; *individual battles and officers*